Designer: Julian Holland

Picture researcher: Susan Brittain

Photo credits:
Aldus Books (with kind permission of Miss M. Strickland); Australian Information Service; Douglas Baglin; BBC Hulton Picture Library; Bodleian Library; Douglas Botting; British Antarctic Survey; British Library Board; British Museum; J. Allan Cash; City Art Gallery, Bristol; Douglas Dickins; Werner Forman Archive; Peter Fraenkel; Garibaldi Museum, Como; Robert Harding Associates; Michael Holford; Kunsthistorisches Museum; London Borough of Richmond upon Thames; Mansell Collection; National Gallery of Art, Washington; National Maritime Museum; National Portrait Gallery; Peter Newark's Western Americana; Operation Drake; Picturepoint Ltd; Axel Poignant; Popperfoto; Rijksmuseum, Amsterdam; G.R. Roberts; Royal Geographical Society; Scott Polar Research Institute; Charles Swithinbank; Tate Gallery; John Topham Picture Library; Zefa Picture Library.

First Published by
Macmillan Children's Books
a division of Macmillan Publishers Limited,
4 Little Essex Street, London WC2R 3LF
and Basingstoke

Adapted and published in the United States by Silver Burdett Company, Morristown, N.J.

1983 Printing

ISBN 0-382-06728-2
Library of Congress
Catalog Card No. 83-50391

Cover: People still explore today. Here members of Operation Drake are transporting stores up the River Ranu, Indonesia, to a camp in the jungle.
Right: Maps improve all the time. Henricus Martellus made this map in 1490 before the east coast of Africa had been properly explored.

The Silver Burdett Color Library

Explorers

Keith Lye

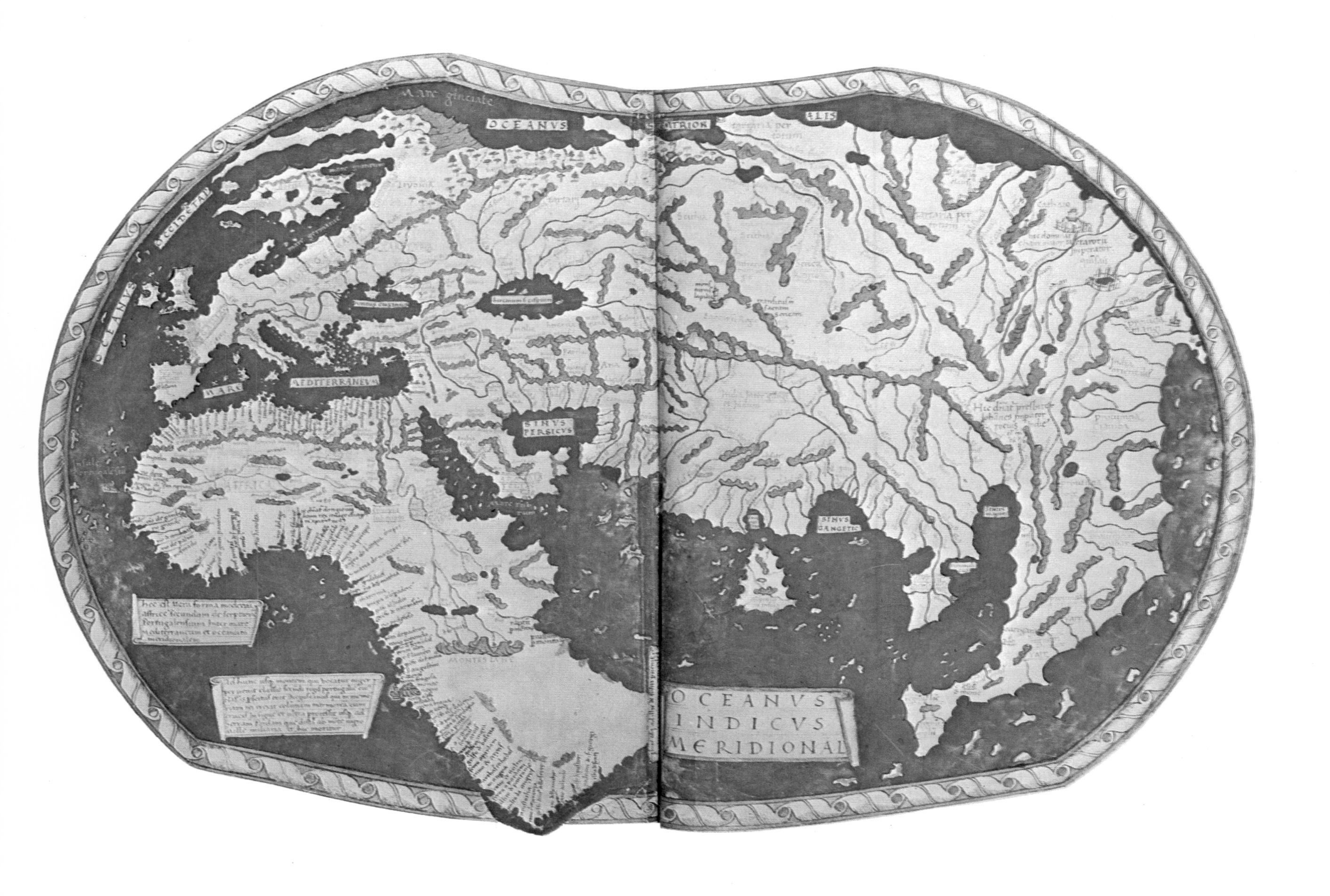

Contents

Right: This calendar was made by the Aztecs, who were American Indians. They founded a great empire in central America from around 1300. The empire was explored and conquered by Spaniards between 1519 and 1521.

Into the Unknown

Today we know what our world looks like. We have seen photographs of the Earth taken from spacecraft. Atlases and maps show us where distant places are. However, even now, there are some areas, such as thick forests, burning-hot deserts and lands covered by ice, which have not been fully explored.

Only 500 years ago, people knew much less about the world. Many people thought that the Earth was flat and that, if they went too far in any direction, they would fall off the edge. Many strange stories were told about unknown lands. Some told of monsters, such as animals with human heads. No such monsters really existed, but occasionally early map-makers put monsters on maps in the areas which they knew nothing about.

Journeys of Adventure

We owe our knowledge of the world to the explorers who went on daring journeys to strange places. People who want to explore must have a love of adventure. But explorers must also gather information about the lands they visit. They must know how to navigate (find their way) and how to measure their position in far-away lands.

Why People Explore

People have many reasons for exploring the unknown. Journeys of exploration were often made to find new peoples to trade with and to discover trading routes. Some journeys were made to find places where colonies could be set up. Explorers were also soldiers. They conquered distant lands and brought back information about them. Others searched for treasure, like gold or precious stones, while some were missionaries who wanted to convert people to their religion. Many explorers were filled only with curiosity and a desire to increase their knowledge of the world.

Right: The Tisisat Falls are near the source of the Blue Nile River in Ethiopia. The search for the source of the Nile, the world's longest river, and its tributaries inspired many explorers.

The Phoenicians lived along the coast of the eastern Mediterranean Sea. They were great sailors. By 1400 BC, they were the main traders in the Mediterranean. They also set up many trading cities around the coast, including Carthage, and later explored the coasts of France and West Africa. A Greek historian wrote of a Phoenician voyage around Africa in about 600 BC.

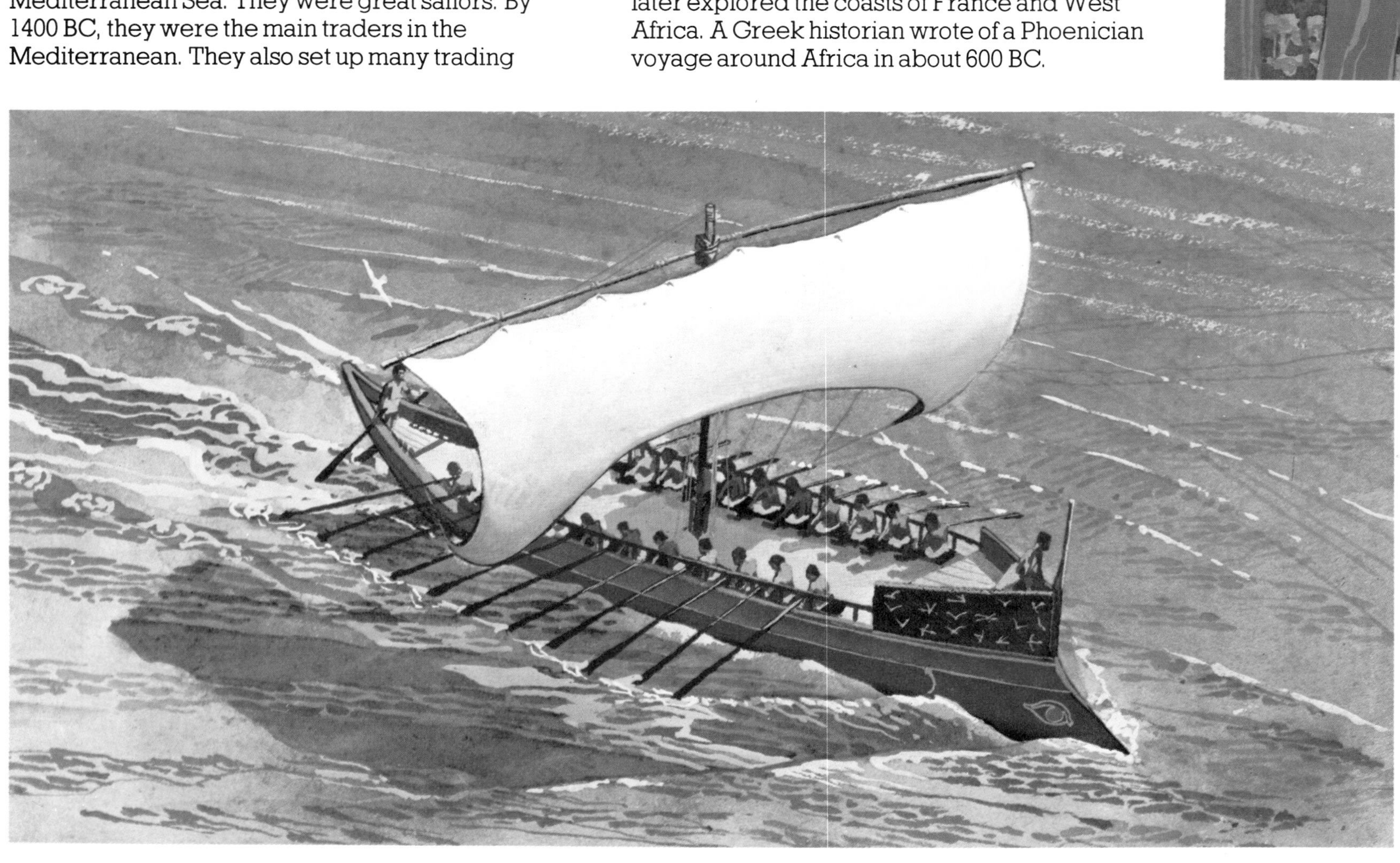

The ancient Greeks had ships which were powered by oars and by the wind. The Greeks set up many colonies. One colony was called Massalia. It stood on the same site as Marseilles, in modern France. A Greek explorer, named Pytheas, sailed from Massalia in about 320 BC. He explored the coasts of Spain and France and sailed around Britain. He may have gone as far north as the coast of Norway.

The map (left) shows some of the important centres of civilization in the ancient world. Many were clustered around the Mediterranean Sea. This sea was the cross-roads of three continents, Europe, Asia and Africa.

A small group of Roman explorers (below) travelled up the River Nile in AD 61. They were sent by Emperor Nero. They did not find the river's source. They probably reached the marshy region in what is now southern Sudan.

The first explorers were prehistoric people who gradually spread around the world. But the first explorers whose names we know were Egyptians. One Egyptian, called Hennu, sailed down the Red Sea in about 2007 BC. He probably reached what is now Somalia. The story of this journey was found carved on a stone.

Egypt faces the Mediterranean Sea, which was the heart of the ancient world. Many civilizations grew up around its shores. One group of Mediterranean people were the Minoans. They had their centre on the island of Crete. Their civilization began around 3000 BC. The people were fine sailors. They explored the Mediterranean on trading voyages.

When the Minoan civilization collapsed in about 1400 BC, the Phoenicians became the leading traders in the Mediterranean. The Phoenicians lived on the eastern coast of the Mediterranean Sea and they were the first to sail out of the Mediterranean into the Atlantic Ocean. The Greek historian Herodotus wrote that a Phoenician ship sailed all the way around Africa in about 600 BC, but no other evidence of this journey has come down to us.

The Phoenicians founded many colonies, including Carthage, in what is now Tunisia. In about 450 BC, a Carthaginian expedition, led by Hanno who may have been king of Carthage, explored the coasts of West Africa. The expedition probably reached what is now Sierra Leone. Hanno saw some strange people, whom he called 'forest dwellers'. These 'people' were probably chimpanzees!

The ancient Greeks greatly increased people's knowledge about the world. The Greeks were restless, curious people. One Greek, Pytheas, sailed around Britain in about 320 BC. Even more important was Alexander the Great who conquered a vast empire that stretched from Greece to India. Alexander was a great soldier and scholar. He took other scholars with him to record his discoveries.

The second great Mediterranean civilization was that of Rome. The Romans destroyed Carthage in 146 BC. They seized a huge empire, but they added little to Greek knowledge of the world.

Early Maps of the World

True explorers always kept records of their travels. They measured the distances they travelled and the positions of places they found, such as islands, towns, mountains and the sources of rivers. When they got home, map-makers used these records to draw maps of the newly-discovered lands. Maps show us what the world, or parts of it, look like. They are not like photographs, but they show us the shapes of continents and countries and the positions of places. From maps, we can work out how far it is between one place and another. We can also measure the compass directions between places.

The first important map-makers were the ancient Greeks. At first, the Greeks thought that the Earth was flat. They believed that there was one land area, surrounded by water on every side. But, more than 300 years before the birth of Jesus, Greek scholars realized that the Earth was a round ball, or sphere. The Greeks could draw proper maps once they understood this.

The Greeks' knowledge of the world was collected together by a scholar named Claudius Ptolemy. Little is known about him, except that he wrote his books in Alexandria, Egypt, around AD 150. Ptolemy wrote one book about astronomy, called the *Almagest.* His second book was called *Geography.* This book explains how to draw accurate maps. It also lists the positions of many places. Ptolemy collected much of his information from a large library which stood then in Alexandria. He also got some information from sailors and traders. But, as far as we know, his book did not contain any maps.

Henricus Martellus made this map in about 1490. It shows the great progress made by Portuguese mariners in the 1400s in

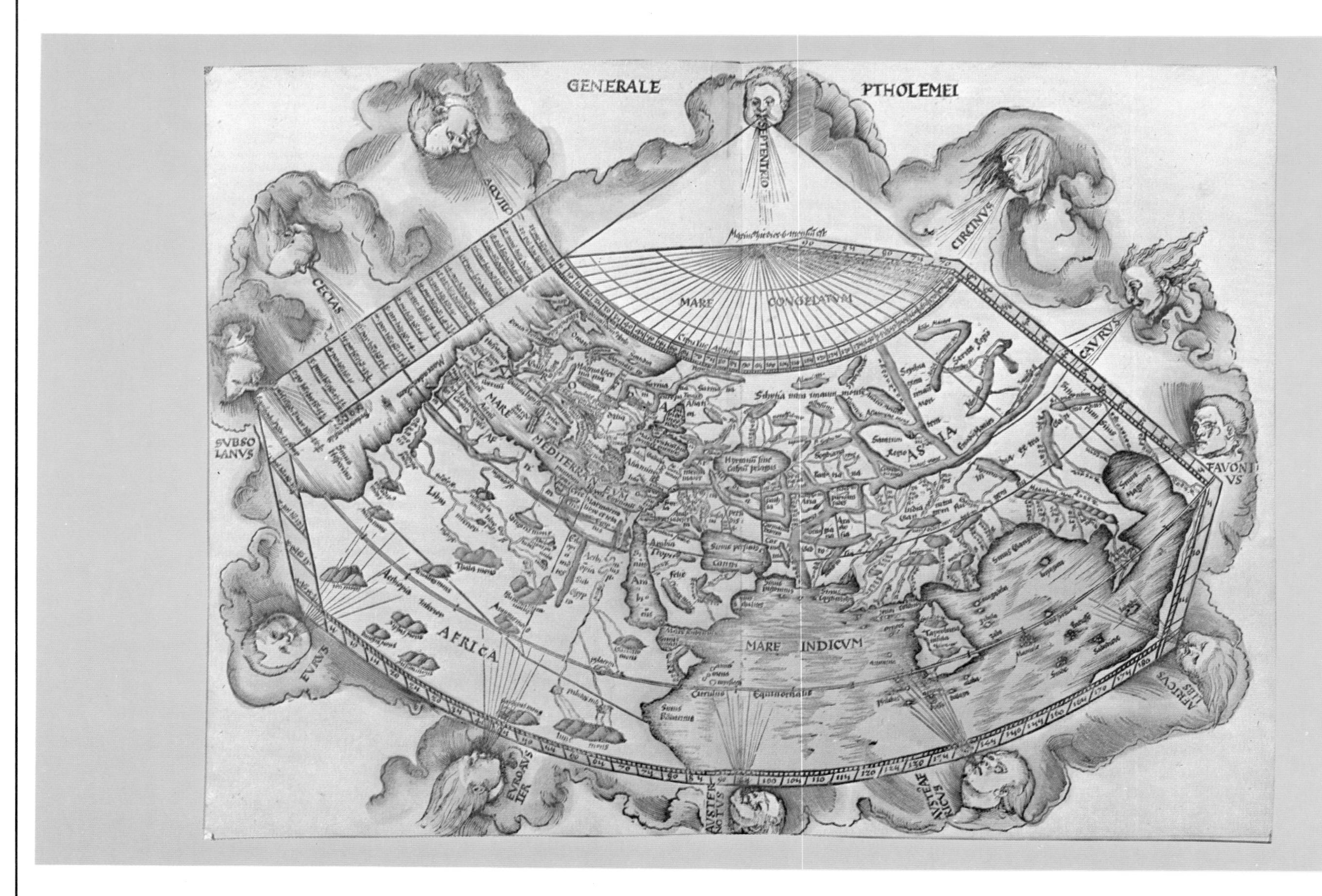

This map was drawn in 1520. It was based on the writings of the Greek geographer, Ptolemy, who lived around AD 150.

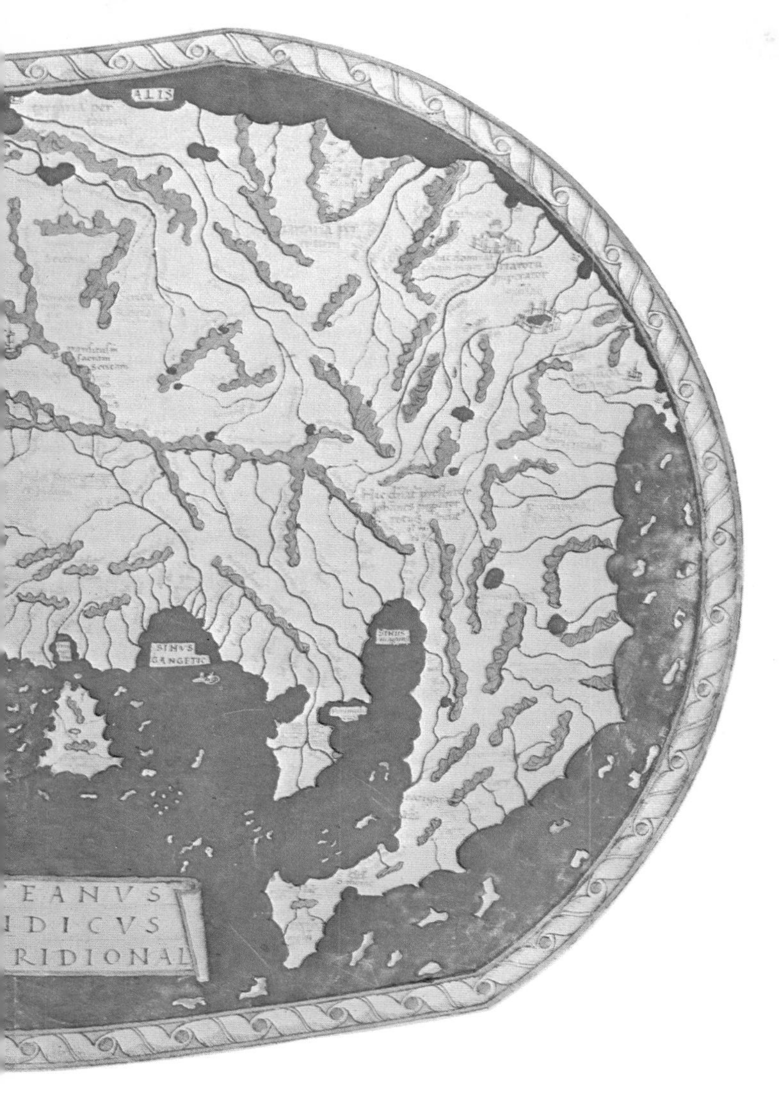

charting the west coast of Africa. Otherwise, it is much like Ptolemy's map (below left).

After Ptolemy died, his books were lost. In the early part of the Middle Ages, many people in Europe went back to the old idea that the Earth was flat. Maps were made, but they were extremely inaccurate.

However, Ptolemy's books were known to the Arabs, who had translated them into Arabic. From the 13th century onwards, various lost works, written by the ancient Greeks, were found in Arab universities and libraries. One by one, they were translated into Italian. Ptolemy's *Geography* was found in 1407. It had a great influence on the explorers who led the Age of Discovery in the 15th century.

Maps were drawn based on the facts that were contained in Ptolemy's *Geography.* One map based on the book is shown on the opposite page. You can see that many things are wrong with the map. For example, Ptolemy thought that India was only a tiny stump of land and that Sri Lanka was much larger than it is. He did not know either that you can sail around Africa.

The map on the left was made by Henricus Martellus in about 1490. He knew that sailors could sail round southern Africa. This had been proved by a Portuguese explorer named Bartolomeu Dias (see page 22). Apart from this, however, his map is similar to the Ptolemy map.

Neither Ptolemy nor Martellus knew of the existence of America. But the map, below, shows the results of the great period of exploration in the 1500s and 1600s when North and South America were charted. By looking at old maps, we can learn how much map-makers knew about the world at the time when they drew the map.

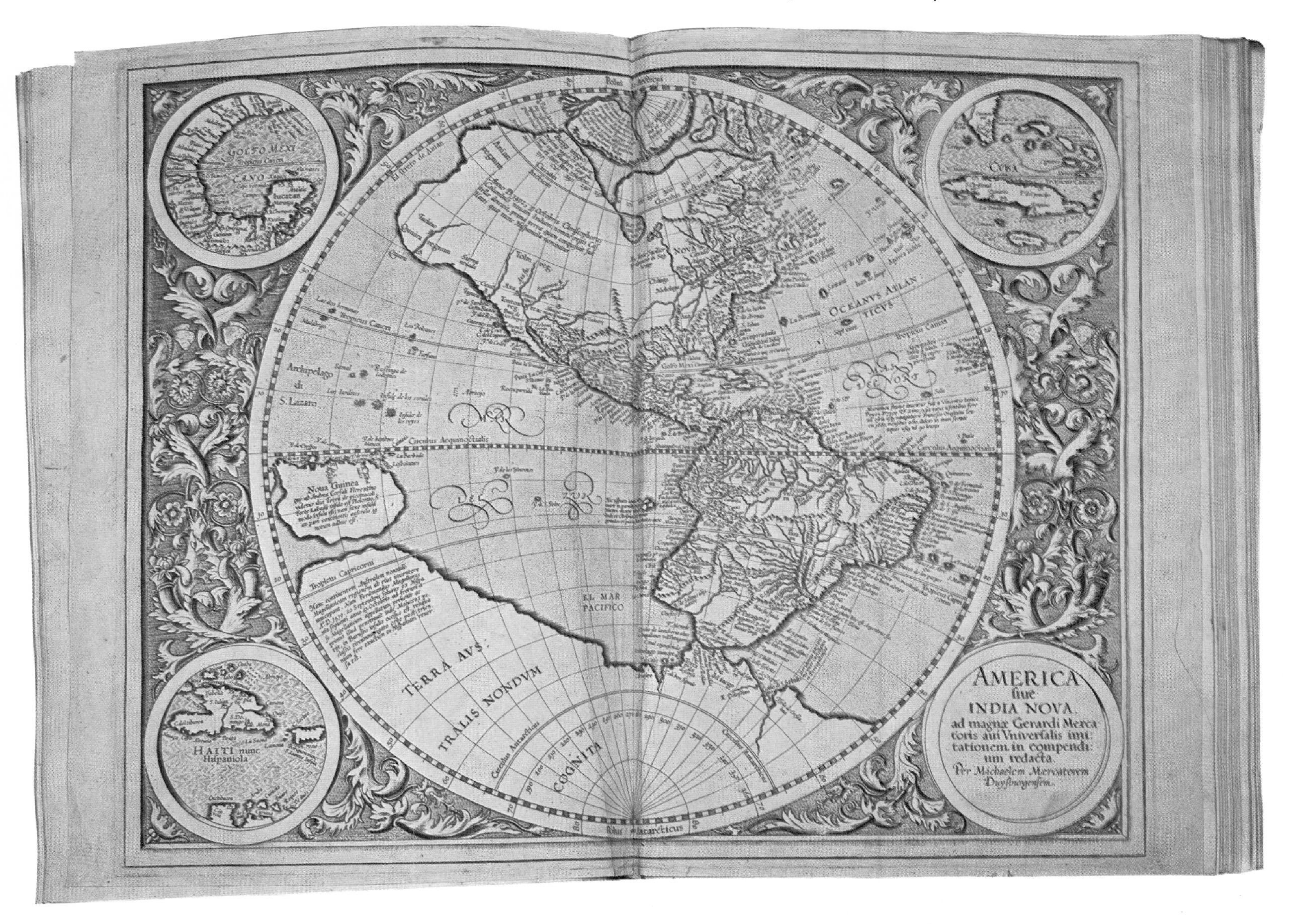

This map of the Americas, published in 1631, was produced by the Flemish geographer Gerhardus Mercator.

Nicolò and Maffeo Polo used horses and camels on their way to China. Nicolò was the father of Marco Polo. Maffeo was Marco's uncle. This picture of the Polos' caravan appeared in the Catalan Atlas of 1375.

The Mongols, under their great emperor Kublai Khan, conquered all of China. The Mongols were the finest soldiers of their day. At its peak, their empire stretched from the River Danube to the China Sea.

Marco Polo was born in Venice in 1254. It was then a rich and powerful city-state. Its power was based on the valuable trade carried on by Venetian merchants between Europe and the Middle East.

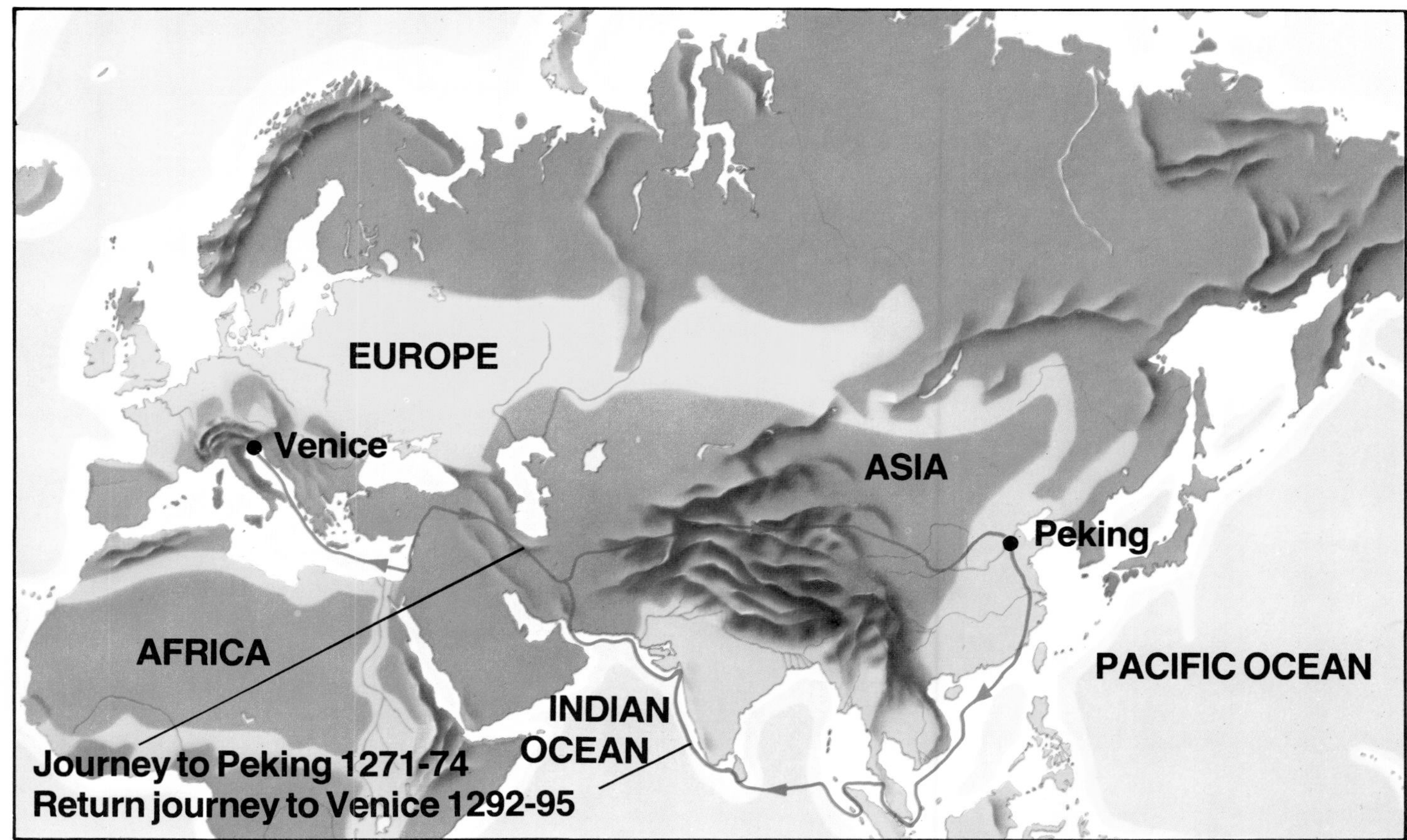

The map shows the route taken by Marco Polo from his home in Venice to the court of the Mongol emperor in China. Marco Polo travelled with his father and his uncle. They faced many dangers on their long journey which began in 1271 and ended in 1274. In 1292, the three men began their long journey home. They travelled partly by sea and partly overland.

In the mid-1200s, the city-state of Venice was an extremely rich trading centre. Venice was the most powerful of several city-states which had grown up in northern Italy. Venetian merchants travelled far and wide. In the lands bordering the eastern Mediterranean Sea, they could buy valuable goods from the Arabs. These had come from China and India. The Venetians themselves did not have trade routes to these far-away lands.

In 1255, two Venetian merchants set out from Venice to Constantinople (now Istanbul in Turkey). These merchants were brothers, Nicolo and Maffeo Polo. They stayed in Constantinople for six years. They then decided to travel and find new trade routes in Russia. They spent three years in the city of Bukhara, which is south-east of the Aral Sea. There, a minor ruler of the Mongol empire welcomed the brothers, who did much business in the area. The Mongol empire had been founded in the AD 1100s and Mongol armies were still conquering new lands. They created an empire which, at its peak, was the largest in the world.

In 1265, some Mongol officials arrived in Bukhara. They were on their way to the court of Kublai Khan at Cambaluc, which was the capital of the empire. Cambaluc stood on the site of Peking, the present capital of China. The officials invited the Polo brothers to go with them. The Venetians agreed and reached Cambaluc in 1266. There they met Kublai Khan.

The Polo brothers returned to Venice in 1268. They began to plan a longer journey and finally set out in 1271. They took Nicolo's son, Marco, with them. Marco was then 17 years old. The journey to China took three and a half years. There were many dangers. Bandits attacked their caravan and they had to cross steep mountains and bleak deserts.

They finally reached Kublai Khan's summer court at Shang-tu, north of Peking. They were warmly greeted by the emperor, who liked young Marco Polo. The emperor sent Marco on many missions, which took him to different parts of China and South-East Asia. Nicolo and Maffeo spent their time in or near the court. But, like Marco, they grew rich because of their friendship with the emperor.

Marco Polo saw more of Asia than any other European until about 600 years later. On his travels, he listed the towns and cities he visited. He noted their size, their products and the importance of trade. He was interested in local customs and wrote about them in his diaries.

In 1292, Marco, together with his father and uncle, began the long journey back to Europe. They reached Venice in 1295. People were amazed by the jewels and beautiful silk robes which the travellers had brought back from China.

Three years later, Marco Polo was captured in a battle between Venice and Genoa. In prison, he told a fellow prisoner of his adventures. The prisoner wrote them down and Marco sent to Venice for his notes and diaries. Finally, Marco's story was published as *The Book of Marco Polo.* This remained the best account of Asia for several hundred years. It inspired such explorers as Christopher Columbus in the 1400s.

The Circumnavigators

After the discovery of the Americas (see pages 14-15), explorers still wanted to find a western route to the Spice Islands (now the Moluccas in Indonesia). Spices from these islands were highly prized in Europe. The man who found a route around South America was a Portuguese nobleman named Ferdinand Magellan. With the support of the king of Spain, Magellan set out with five ships on September 1519. In early 1520, he explored the coast of South America. He was slowed down by bad weather and one of his ships was wrecked. He reached the mouth of what we now call the Strait of Magellan in mid-October. While he was exploring this strait, one of his ships deserted and went home.

Ferdinand Magellan led the first voyage around the world and he was the first man to find a route around South America. But he was killed in the Philippines.

The Strait of Magellan links the Atlantic Ocean to the Pacific Ocean. It extends between the mainland of South America and the islands of Tierra del Fuego. It was discovered by Magellan and named after him.

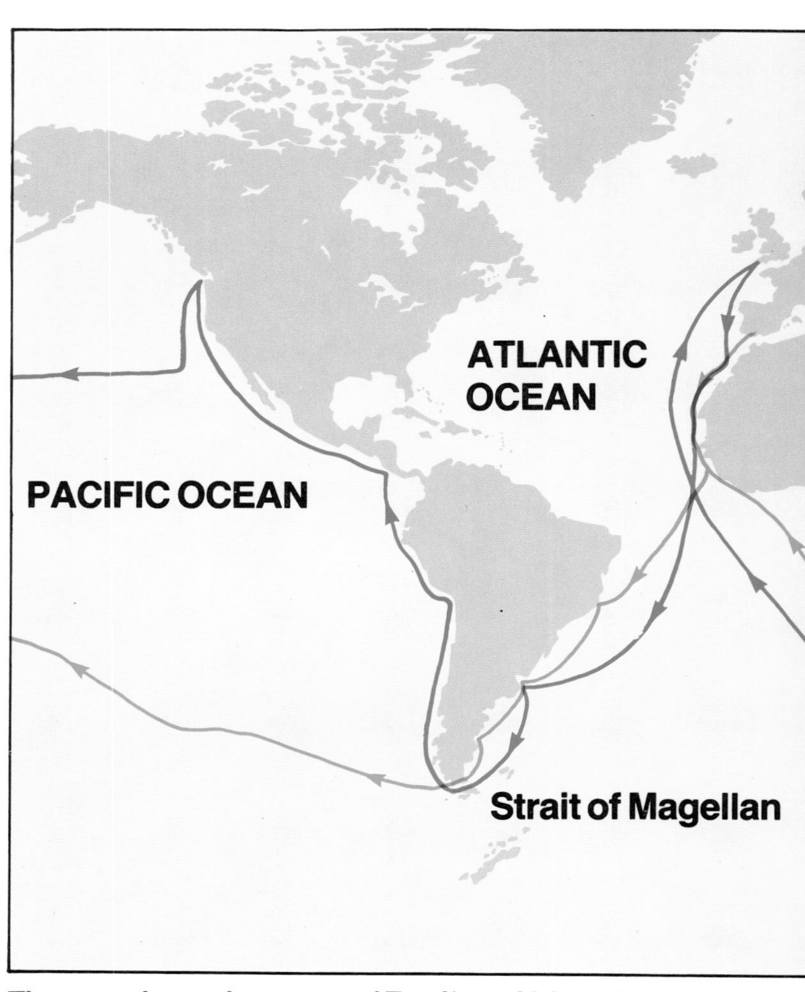

The map shows the routes of Ferdinand Magellan and Sir Francis Drake on their great voyages around the world. Magellan was a Portuguese nobleman in the service of Charles V of Spain. He achieved what Columbus had

Ferdinand Magellan's ships took 38 days to pass through the Strait of Magellan, although it is only about 560 kilometres long. Magellan went ahead of his ships in small boats. This helped him to find the best routes for his ships. At

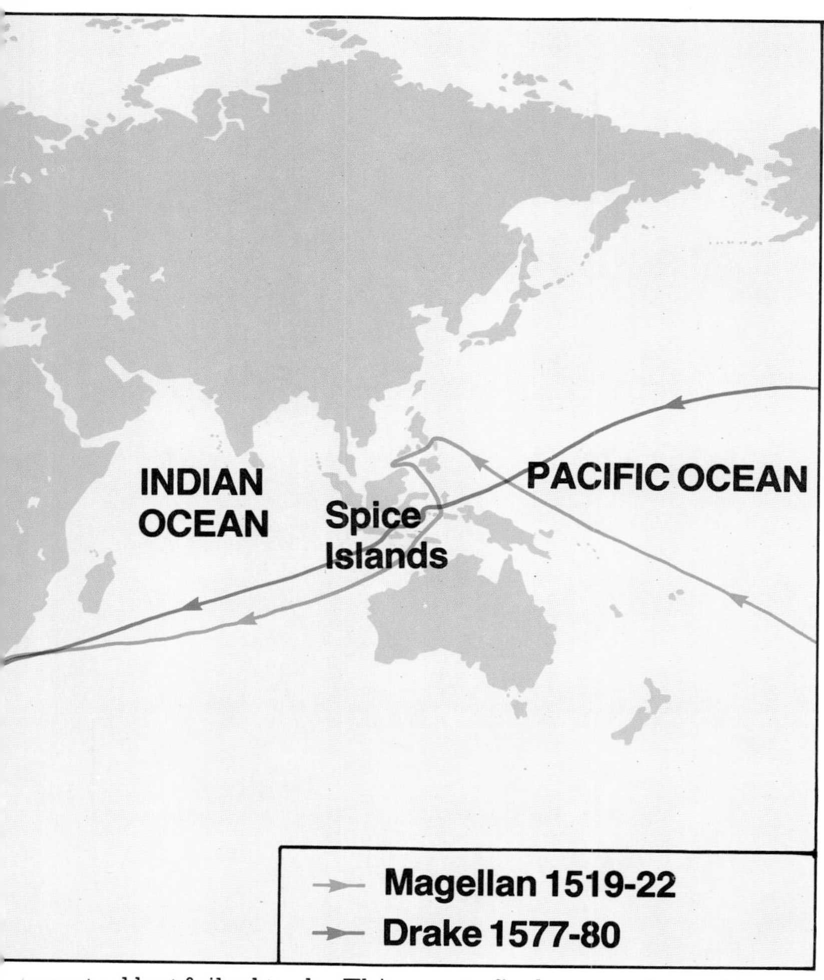

wanted but failed to do. This was to find a way of reaching the Spice Islands by sailing west. The first ship to sail around the world was the *Victoria*. It was one of the five that set out in 1519.

this time, one of his ships deserted and sailed back to Spain. Another ship had already been wrecked and so Magellan was left with only three ships when he finally reached the Pacific Ocean.

On the long voyage across the Pacific, food and water were in short supply and many men became ill. Land was first sighted in late January 1521. In March, the ships reached the Philippines, where Magellan was killed in a battle on April 27. The ships continued on to the Spice Islands. One ship, the *Victoria,* reached Portugal in September 1522. It was captained by Juan Sebastian del Cano. This voyage greatly increased people's knowledge of the world.

Sir Francis Drake was the next man to circumnavigate the world. He set out in December 1577, with three ships. But, before reaching the Pacific Ocean, one ship was wrecked and another had turned back. Drake was left in command of the *Pelican,* which he renamed the *Golden Hind.* When rounding South America, Drake probably discovered Cape Horn, the tip of South America. In the South Pacific, he seized treasure from Spanish settlements in Chile. He also claimed California for Britain. Drake sailed on to the Spice Islands, India and around Africa, before reaching England in September 1580. He had added little to people's knowledge of the world, but his voyage was a great achievement.

Sir Francis Drake, the most daring seaman of the Elizabethan age, was the second man to sail around the world. He returned with enormous treasures which he had seized in raids on Spanish settlements in Chile.

The First European in America

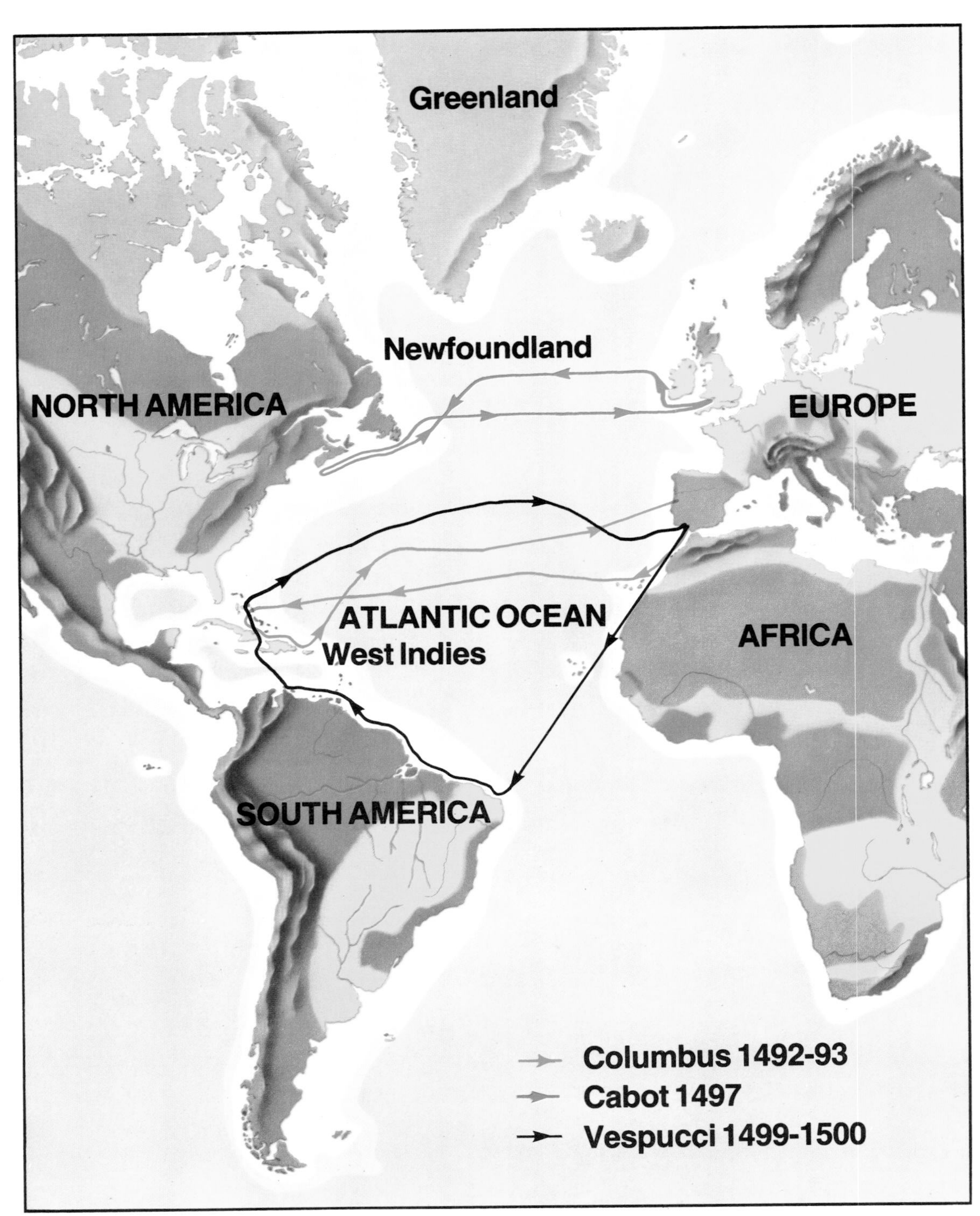

The map shows the routes of three great voyages of discovery. They were made in the 1490s by Christopher Columbus, John Cabot and Amerigo Vespucci. America comes from Vespucci's first name.

This Indian village was painted in 1587 by an English colonist. The people of the Americas were called Indians because early explorers thought they had found Asia.

John Cabot, an Italian, led an English expedition from Bristol across the Atlantic Ocean. He had set out to find China. Instead, he found the coast of Canada.

The first European (left) to land in North America may have been a Viking, Leif Ericson. Ericson sailed from a Viking colony in southern Greenland and probablylanded in north-eastern North America shortly after AD 1000.

Christopher Columbus (inset right) sailed westwards across the Atlantic to find a new route to the Spice Islands in Asia. He landed in the West Indies, but he thought that he was in Asia. He made this mistake because he thought that the Earth was much smaller than it really is. He got this wrong idea from the book *Geography,* by the Greek scholar, Claudius Ptolemy. Ptolemy had underestimated the circular distance around the world by nearly two-fifths. Columbus was born in Genoa, Italy.

North and South America make up nearly three-tenths of the land on Earth. But, until 1492, they were unknown to people in Europe.

Yet Europeans had probably landed in North America many years earlier. These people were Viking sea-rovers from Scandinavia. The Vikings set up colonies in Iceland and Greenland. Soon after AD 1000, a Viking, Leif Ericson, sailed from Greenland down the Canadian coast. He probably landed in what is now the north-eastern United States. He called the place Vinland, or 'Wineland'. This is because he found grapes growing there.

No one knows whether Christopher Columbus had heard of Ericson. But Columbus believed that the Earth was round. (Most people at this time thought the Earth was flat.) He was convinced that he could reach Asia by sailing west. He thought that it might be quicker than the eastern route.

Backed by Spain, Columbus set out on 3 August 1492. He had three ships: his flagship, the *Santa Maria,* and the smaller *Niña* and *Pinta.* Some of his men feared that they were going to their death. They were all relieved when land was sighted on October 12. They had reached the Bahamas, a West Indian island group. They sailed on and discovered Hispaniola (which is now Haiti and the Dominican Republic). Columbus returned to Spain in triumph but unaware that he had discovered a new continent. He later made three other voyages to the Americas. But he believed, right up to his death, that he had reached Asia. He thought that the Bahamas were part of the 'Indies', near China. This is why he called the people he found living in the Americas 'Indians'.

In 1497, an Italian, John Cabot, led an English expedition across the North Atlantic. Like Columbus, he was looking for Asia, but he found the Canadian coast. He, too, did not understand that he had discovered part of a new continent.

On his third voyage, Columbus sighted South America. The northern coast of South America was explored in 1499 and 1500 by an Italian merchant, Amerigo Vespucci. Vespucci became convinced that this was not Asia, but a 'New World'. Gradually, more and more people came to agree with him and the word America was made up from his name Amerigo.

Cabot's son, Sebastian, went with his father on the voyage of 1497. He later became an explorer in his own right.

Jacques Cartier explored the St Lawrence River in the 1530s and 1540s. Samuel de Champlain reached the Great Lakes in 1615. The map also shows the route taken by La Salle down the River Mississippi in the 1680s.

This map, drawn by a Portuguese map-maker before 1547, looks odd because north is at the bottom. Turn it upside down and you will see the St Lawrence.

Samuel de Champlain was a French explorer who founded the city of Quebec. His discoveries encouraged people to settle in Canada.

Indian Wars were one of the dangers faced by explorers in Canada. Samuel de Champlain became friendly with the Huron Indians. In 1609, Champlain helped the Hurons in a raid on their enemies, the Iroquois. Champlain, with two of his own men, used firearms to frighten the Iroquois who fled. Afterwards, the

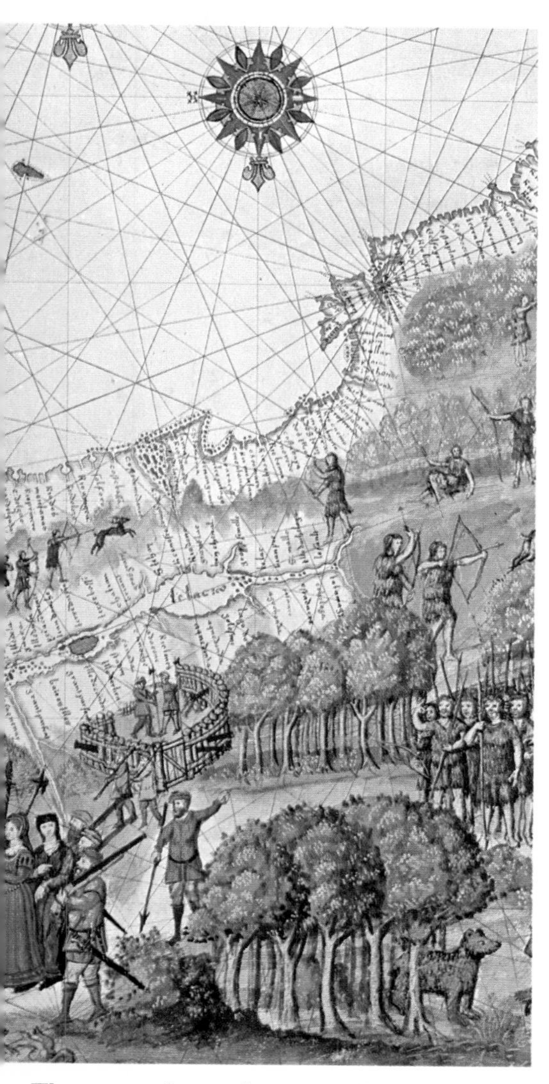

The map shows Jacques Cartier arriving in Canada with French settlers and American Indians.

This painting shows the departure of Robert Cavelier, Sieur de la Salle, from Fort Frontenac on Lake Ontario in 1678. He sailed through lakes Erie, Huron and Michigan. Later, La Salle explored the Mississippi valley and claimed it for France. He named it Louisiana after King Louis XIV.

Exploration of North America continued in the early 1500s. Spaniards explored the southern parts of North America. In 1513, an elderly Spaniard, Juan Ponce de Léon, discovered Florida. He was looking for a magic mountain where men were supposed to be made young again.

An officer, Alvar Núñez Cabeza de Vaca, explored Florida and was the first European to see the River Mississippi. A Spanish adventurer, Hernando de Soto searched for gold in what is now the south-eastern United States. Another, Francisco de Coronado, travelled from Mexico through the south-west, where he saw the Grand Canyon in 1540.

In northern North America, seamen sought a way through or around the continent. North America was not a great discovery to them. It was simply a nuisance which blocked their way to the Spice Islands and India.

John Cabot's voyage of 1497 was followed by another in 1498. On this second journey, he explored about half of the coast of North America. In 1509, his son, Sebastian, sailed into the Hudson Strait. He may have entered Hudson Bay, but his crew made him return to Bristol. The British were disappointed with the explorers, because no one had found a North-West Passage to Asia (see pages 38-39).

Early exploration of the interior of Canada was undertaken mainly by Frenchmen. Soon, Canada was spoken of as 'New France'. In 1534, a French seaman, Jacques Cartier, led an expedition to Canada. He explored the mouth of the St Lawrence River. In 1535, he sailed up this wide river. Cartier's journeys had greatly increased people's knowledge of North America. But he had also proved that the St Lawrence River was not the North-West Passage through America to Asia.

French interest in Canada revived in the early 1600s. This was caused largely by the efforts of the explorer Samuel de Champlain. He visited Canada for the first time in 1603. In 1608, he founded a settlement that grew into the city of Quebec. In 1615, he explored Lake Ontario and Lake Huron. In 1620, a friend of his, Etienne Brulé, reached Lake Superior. He had discovered the world's largest fresh water lake.

Some people who went to Canada were missionaries, like Father Jacques Marquette. Marquette and a fur trader, Louis Joliet, explored the Mississippi region south of the Great Lakes in 1673. Their work was followed up by Robert Cavelier, Sieur de La Salle.
La Salle, an educated man, went to Canada to make his fortune. He became a fur trader, but he made several great journeys of exploration in the Great Lakes and the River Ohio regions. He also made a journey down the River Mississippi to the sea.

Iroquois regarded the French as their enemies. In the 1700s, they fought with the British against the French.

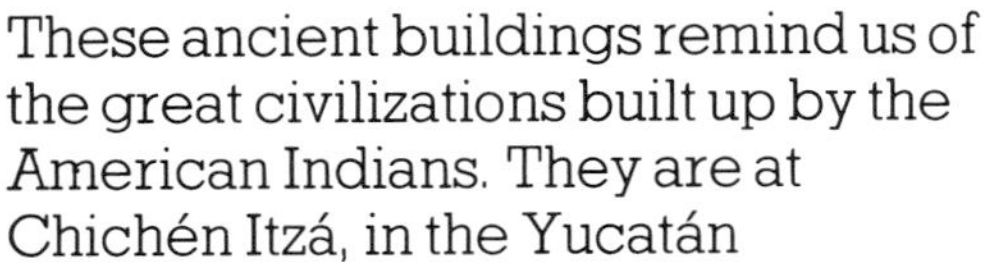

These ancient buildings remind us of the great civilizations built up by the American Indians. They are at Chichén Itzá, in the Yucatán peninsula of Mexico. Chichén Itzá was founded by the Maya Indians who flourished between AD 600 and 900. It was later enlarged.

The Aztec city of Tenochtitlán was captured and destroyed by Hernando Cortes in 1521. The Aztecs had never seen firearms or horses.

The American Indians of North, Central and South America varied greatly. Some lived simple lives as hunters. Others founded huge civilizations. One early civilization in Central America was founded by the Mayan Indians. But it was weak when the first Europeans arrived. The most powerful people in Central America at that time were the Aztecs of central Mexico. And, in South America, there was the great empire of the Incas.

The first European to reach Mexico was Francisco Fernández de Córdoba in 1517. A second expedition in 1518 learned of the Aztecs and their enormous wealth in gold.

In 1519, a Spanish adventurer, Hernando Cortes, led a small army against the Aztecs and their emperor, Montezuma II. The Aztecs were afraid of the Spaniards' guns and horses, neither of which they had ever seen before. Montezuma did not want to harm Cortes because he looked like an Aztec god, who was expected to appear in 1519. Cortes captured Montezuma and killed him. In 1521, Cortes destroyed the Aztec capital, Tenochtitlán, and ended Aztec power.

Another Spaniard, Francisco Pizarro, led an army against the Incas in 1531. In 1532, Pizarro captured the Inca emperor, Atahualpa, and demanded a great ransom of treasure for his freedom. But even though the ransom was paid, the Spaniards killed Atahualpa. The Incas were finally crushed in 1569.

The conquests of Cortes and Pizarro in the Americas were cruel and brutal. But they greatly increased Europe's knowledge of the Americas and opened up new lands to Europeans who wanted to settle.

Gold objects, like this mask, were among the many treasures sought by the Spaniards in Peru.

Francisco Pizarro conquered the rich Inca empire with a force of only 180 men. He tricked the emperor Atahualpa

Tenochtitlán stood on an island in a large lake. Three causeways linked the city to the mainland. The Spaniards called it 'the Venice of the New World'. The lake has now gone and Mexico City stands on the same site.

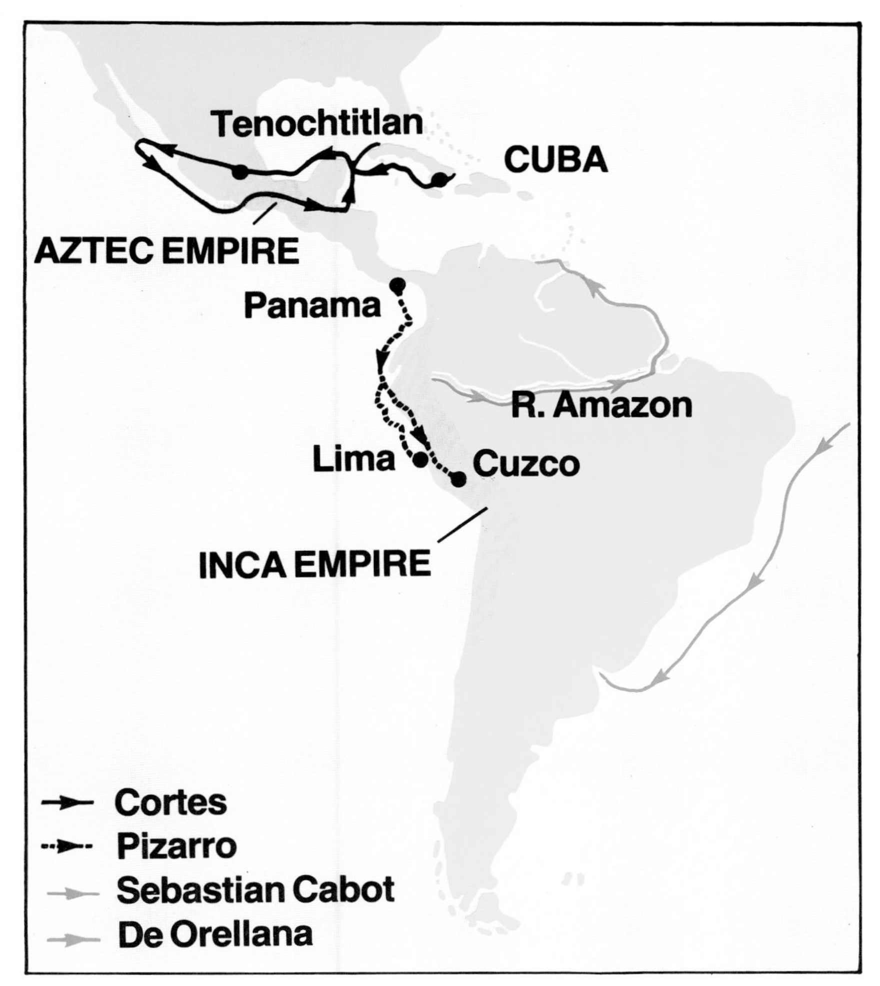

The map shows four important journeys of discovery. Those made by Hernando Cortes and Francisco Pizarro were military conquests in search of treasure. Sebastian Cabot explored the River Plate region for Spain in 1526. The Spaniard Francisco de Orellana served under Pizarro in Peru. He then went east to explore the River Amazon. He followed the river to its mouth in the Atlantic Ocean.

into coming to meet him. When the emperor arrived, Pizarro's men seized him and killed many of his bodyguards.

After the conquest of Peru, a Spaniard, Diego de Almagro, went on a journey of exploration to the south. On his way to Chile, he saw Lake Titicaca. This is the highest navigable lake in the world, it now forms part of the border between Peru and Bolivia. Indian boats on the lake are made of reeds.

The Grand Canyon is one of the World's natural wonders. It is in the south-western part of the United States. It was discovered by a Spanish expedition in 1540. The mission, led by Francisco de Coronado, was searching for treasure, but it was unsuccessful.

Fur trappers, hunting for beaver, blazed trails all over western North America. Some of them made important contributions to the exploration of unknown lands.

While crossing the United States to the Pacific Ocean in 1804-06, Captain Meriwether Lewis and William Clark contacted Indian tribes who were unknown at that time.

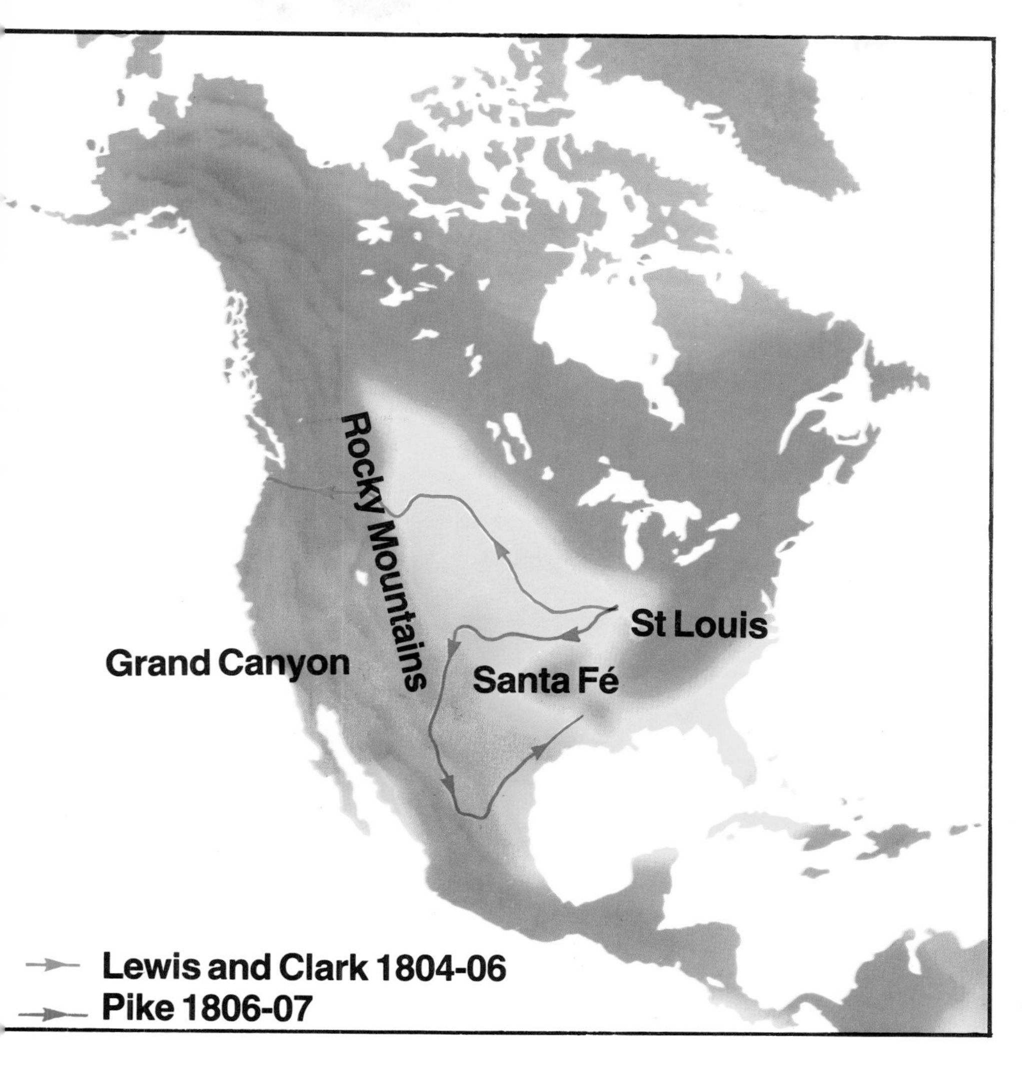

John Charles Frémont (above) was an American soldier, who explored and mapped areas between the Rockies and the Pacific Ocean coast in the 1840s and 1850s.

The map (left) shows the routes of two important journeys of exploration in the rugged country of western North America.

Zebulon Pike, an American soldier, explored the south-west in 1806-7. He entered Spanish territory and was arrested. He was taken to Santa Fé (above) and returned to the United States.

In 1783 the 13 British colonies in eastern North America became the United States. This meant that all of North America was claimed either by the United States, Britain or Spain. But large areas were still unexplored.

In Canada, a Scottish fur trader and explorer, Alexander Mackenzie, believed that he could find a way to the Pacific Ocean. In 1789, he set out from a fort on Lake Athabasca, in north-central Canada. He went northwards to the Great Slave Lake and followed a mighty river which flowed from it. This river, which is now called the River Mackenzie, led him to the Arctic Ocean. This was a great discovery, but Mackenzie was disappointed that it was not the Pacific. In 1792, he set out again. He crossed the Rocky Mountains and eventually reached the Pacific Ocean. He had made the first overland crossing of North America.

The first overland expedition across the United States was led by Captain Meriwether Lewis and his friend, a former soldier, William Clark. They set out from St Louis in May 1804 and reached the Pacific Ocean in the north-western United States in December 1805. They got back to St Louis in September 1806. They had faced many hardships, but they brought back important records of their explorations. They also brought reports of many Indian tribes who were unknown before then.

Other explorers, like Zebulon Pike and John Charles Frémont, explored the south-western parts of the United States. The new routes they found were used by settlers who were spreading gradually across North America.

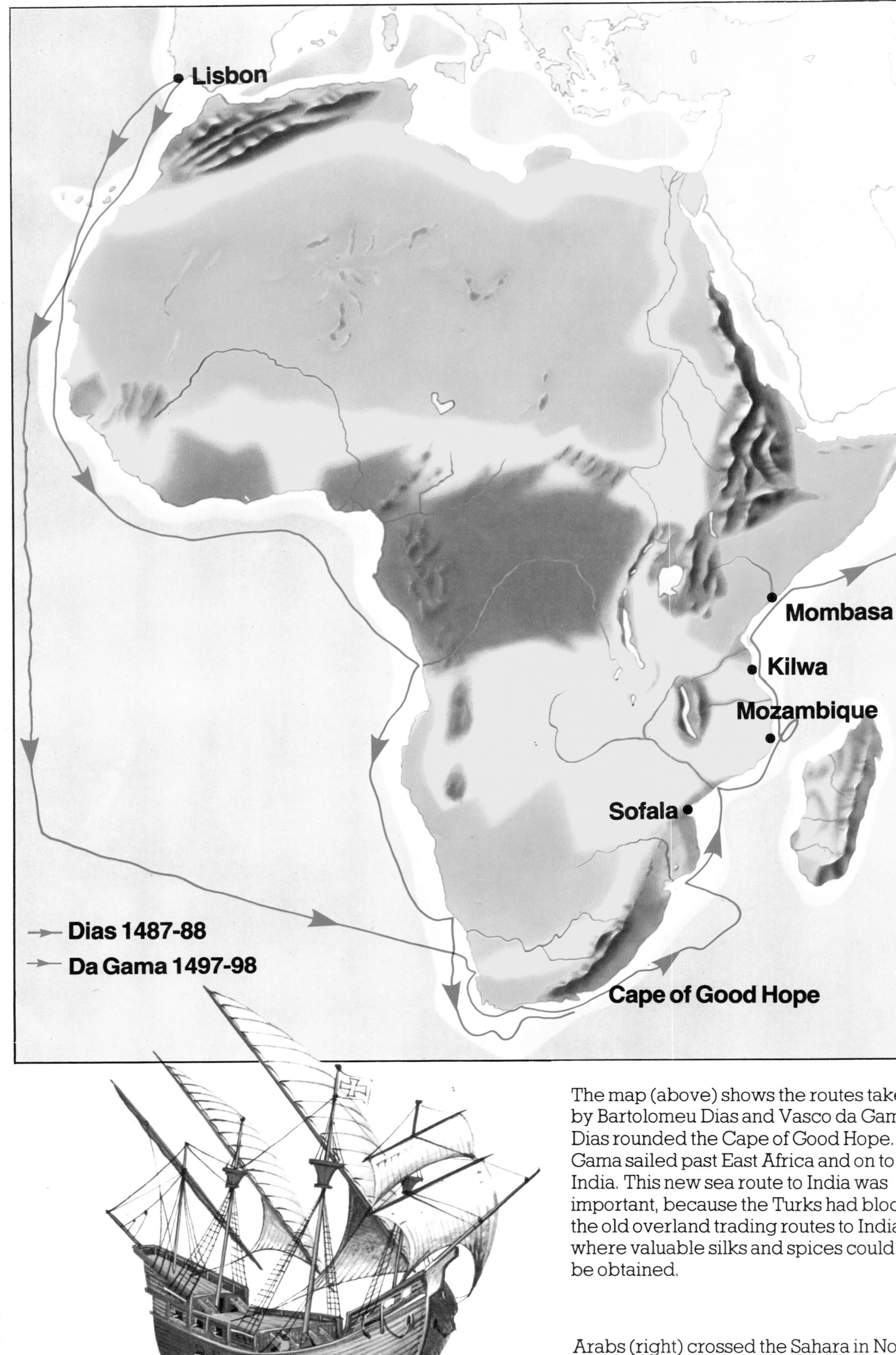

The map (above) shows the routes taken by Bartolomeu Dias and Vasco da Gama. Dias rounded the Cape of Good Hope. Da Gama sailed past East Africa and on to India. This new sea route to India was important, because the Turks had blocked the old overland trading routes to India, where valuable silks and spices could be obtained.

Caravels were ocean-going ships developed in the 1400s in Portugal and Spain. Explorers sailed in them. The square sails were used when strong winds were blowing from behind the ship. The triangular sails were used when the winds were light or blowing from the side.

Arabs (right) crossed the Sahara in North Africa regularly in the Middle Ages in camel caravans. They traded with medieval kingdoms in black Africa. Most of our knowledge of these early kingdoms, such as ancient Ghana and ancient Mali, comes from accounts written by Arabs. But, at this time, the people in Europe knew nothing of Africa south of the Sahara.

Prince Henry of Portugal (left) was called 'the Navigator'. He ran a school of navigation in Portugal and sponsored many expeditions to explore the coasts of Africa. But he, himself, never went on any of the expeditions.

The European Age of Exploration began in the 15th century. One reason why Europeans needed to explore was that they wanted to find new ways to Asia. New trading routes were needed because, after 1453, the powerful Turkish empire had blocked the old routes.

One man led the way in the early part of the Age of Exploration. He was the third son of the king of Portugal, Prince Henry. Henry was a man of great learning. He set up a school of navigation at Sagres, in southern Portugal. He invited travellers and scholars to his school to build up a store of knowledge about the world. He also got ship-builders to design a new kind of ship, called the caravel. This ship was designed for long ocean voyages.

Prince Henry sent many expeditions to map the western coast of Africa. This work continued after his death in 1460. In 1487, a Portuguese sea captain, Bartolomeu Dias, sailed down the coast of Africa, far beyond the parts which had already been explored. Then a storm blew his caravels southwards. When the storm ended, he could not see land. He sailed east, expecting to find the African coast. But he saw nothing. Finally, he realized that he must have gone past the southern tip of Africa. He turned north and, at last, found the coast of southern Africa. Dias wanted to go on, but his tired crew made him turn back.

In 1497 and 1498, another Portuguese mariner, Vasco da Gama, achieved what Dias had wanted to do. Despite many hardships, da Gama sailed around southern Africa and up the coast of East Africa. There, he visited various Arab trading centres that had been flourishing for hundreds of years. He then sailed across the Indian Ocean to Calicut, in southern India. He returned home in triumph. His discovery made many Portuguese merchants very rich.

The Portuguese set up forts (left) on the coast of East Africa. They wanted to take over the trade which was in the hands of the Arabs. This is Fort Jesus, which was built in the 1590s at Mombasa, in Kenya.

Exploring the Niger

There was little interest in the interior of Africa until around 1800. Forts and trading posts had been built on the coast, where European traders bought goods like gold, ivory and slaves. (The slave trade went on from the 1400s until the 1800s.) European traders usually bought slaves and other products on the coast. The African chiefs on the coast were supplied with these goods from inland areas.

Africa seemed an unfriendly place to European explorers. In some areas, dense coastal forests blocked the way of travellers. Many rivers had rapids and waterfalls which prevented ships sailing up them. There were also hostile Africans, wild animals and dangerous diseases which were unknown to European doctors.

In the late 1700s, learned societies in Europe started to take an interest in Africa. One was called the African Association. It was formed in London in 1788 by Sir Joseph Banks. He had explored the Pacific Ocean with Captain Cook (see pages 34-35). The Association was interested to learn about two things. The first was the course of the River Niger. The second was the mysterious city of Timbuktu which is near the river.

In 1794, the Association hired Mungo Park, a young Scottish doctor, to explore the River Niger. In 1795, he left what is now Gambia and reached the River Niger at Ségou, in what is now Mali. He travelled down the river for three days, but had to return because he was tired and ill. He got back to Britain in December 1797.

In 1805, Park returned to Africa. He led a large party, including 31 British soldiers. But, as they marched inland, his men started to fall ill. Only ten men were alive when he reached the River Niger. The survivors went down the Niger by boat. They reached the falls at Bussa, in what is now Nigeria. There, they were drowned as they tried to escape an attack by a number of hostile African warriors.

Some people thought that the Niger flowed into Lake Chad so a British team went to explore there to see if this was true. The team consisted of Walter Oudney, Hugh Clapperton and Dixon Denham. In 1822, they crossed the Sahara from the north and became the first Europeans to see the lake. Clapperton and Oudney wanted to continue their journey and find the River Niger, but Oudney died. So Clapperton went back east, and on to North Africa with Denham.

Clapperton was back in Africa in November 1825. One of his party was Richard Lander, his manservant. The party reached Bussa where Park had died earlier.

Kano, in northern Nigeria, was an important city on the Saharan caravan routes. A Scottish naval lieutenant, Hugh Clapperton, visited Kano in the 1820s. He recorded local customs.

This picture of Timbuktu, an African city south of the Sahara, was drawn by a French explorer, René Caillié, in 1828. Until the 1820s, Timbuktu had not been seen by any European. But the Arabs had been trading there for many centuries. An Arab geographer, Ibn Batuta, had visited it in the 1350s.

Richard Lander and his brother John travelled down the River Niger by canoe in 1830. They met some hostile Africans on their way

Elmina was founded in 1482 on the coast of what is now Ghana. It was the first of several forts built on the coast of West Africa by the Portuguese, who had found that the region was rich in gold. The forts were meant to stop other Europeans from getting the gold.

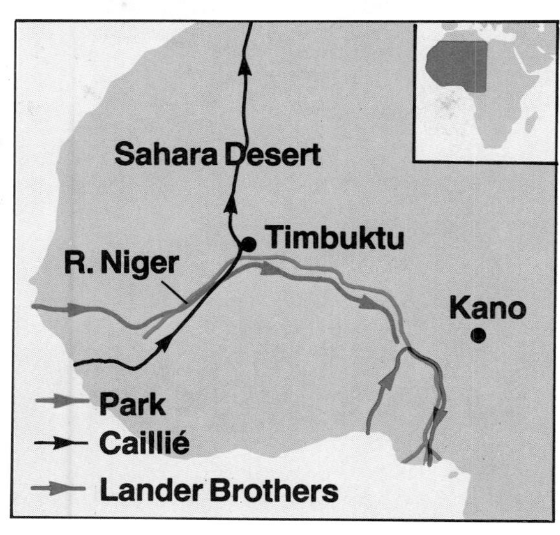

The map shows three journeys of exploration in north-western Africa. The course of the River Niger, one of Africa's many mysteries was partly charted by Mungo Park and partly by Richard and John Lander.

Mungo Park, a Scot, made two journeys to explore the upper part of the River Niger. He was drowned in 1806, when trying to escape an African attack.

The Lander brothers set out from Bussa in Nigeria, because that was the place on the River Niger where Mungo Park was drowned. Richard Lander had already been to Bussa when he was manservant to Hugh Clapperton. When Clapperton died in 1827, Lander went back to Britain.

One by one, the members of the expedition died. Lander was the only survivor. He went back to Britain in 1827. In 1830, he and his brother returned to Africa and traced the River Niger's course from Bussa to the sea.

The prize of being the first European to reach Timbuktu was achieved by a Scot, Major Alexander Laing in 1826. But Laing was murdered near the city shortly after arriving.

In 1828, a French explorer, René Caillié, reached Timbuktu. He was disappointed to find that it was no longer a wealthy place. Caillié stayed two weeks and then crossed the Sahara to Morocco. In Paris, he was awarded 10,000 francs for his achievement.

David Livingstone was a Scottish doctor, who became a great missionary-explorer in southern and east-central Africa. He saw the terrible effects of the slave trade and exposed its horrors to the world.

The famous explorer, David Livingstone, was born in 1813 near Glasgow in Scotland. His family was poor and he started work at a cotton mill at the age of ten. But he taught himself many subjects and became a doctor of medicine. In 1840, he went to South Africa as a missionary with the London Missionary Society.

In Africa, he wanted to spread the Christian message as widely as possible. This led him to make journeys of exploration. In 1851, he travelled north and found a large river. He thought, rightly, that this must be the River Zambezi. In 1852 he explored the upper Zambezi region. He then travelled, usually on the back of an ox, to the sea in Angola. He returned to the upper Zambezi and followed it eastwards to the sea. He discovered the Victoria Falls on this journey. He reached the mouth of the Zambezi in May 1856.

On his travels, Livingstone learned much about the slave trade. He had seen deserted villages and many people killed by slave traders. He returned to Africa in 1858 as a British consul in Mozambique to oppose the slave trade. He also commanded an expedition to explore eastern and central Africa. He explored the Lake Nyasa region. (Lake Nyasa is now also called Lake Malawi.)

In 1866, Livingstone began his last great journey. In searching for the source of the River Nile, he visited Lakes Mweru and Bangweulu, but neither was the source of the Nile. On his travels, he ran out of supplies and fell ill. Henry Morton Stanley brought him supplies in 1871. After Stanley left, Livingstone set off south again. He died in 1873. His servants buried his heart and other internal organs under a tree. They took his body to the coast. Because of their love for him, his body now lies in Westminster Abbey, London.

Slaves were captured in the interior. They were yoked together and driven to the east coast. They were finally sold at the slave market on the island of Zanzibar. Many slaves died before reaching Zanzibar.

Local people called these falls 'the smoke that thunders'. Livingstone named them the Victoria Falls on 17 November 1855. He was the first European to see the falls.

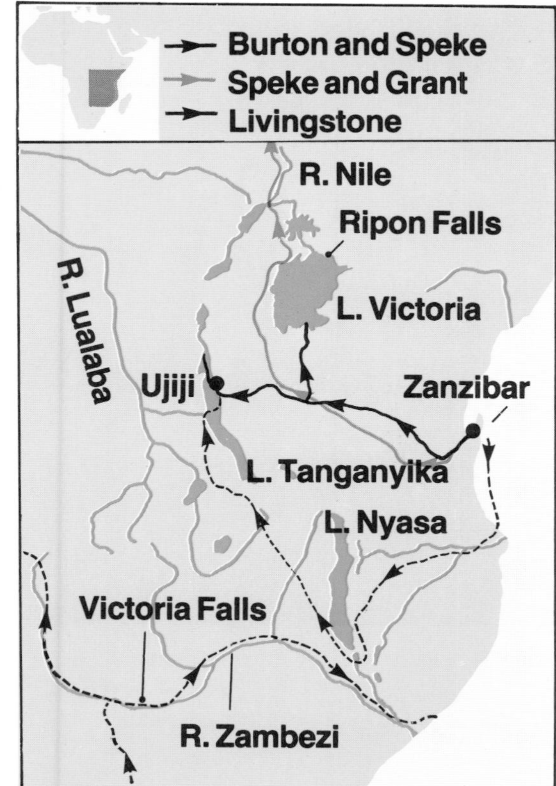

The map shows some of the great journeys made by explorers in eastern and southern Africa to chart the rivers and lakes of the region.

Henry Morton Stanley (left) found David Livingstone on 28 October 1871, at Ujiji, a port of Lake Tanganyika. He brought much-needed supplies to Livingstone, who had been weakened by lack of food and medicines.

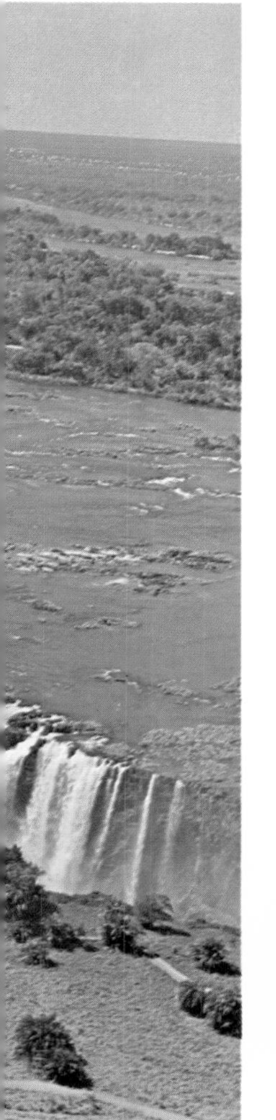

John Hanning Speke is shown here against a background of the Ripon Falls, the northern outlet of Lake Victoria. When Speke first saw Lake Victoria in 1858, he was convinced that it was the source of the River Nile.

The River Nile attracted many explorers before David Livingstone. A Scot, James Bruce, had explored the Blue Nile in 1770. But no one knew where the main river, the White Nile, started. In 1856, two explorers, Richard Burton and John Hanning Speke, set out to search for the source of the White Nile.

Burton and Speke marched inland across East Africa. In 1858, they became the first Europeans to see Lake Tanganyika. When Burton fell ill, Speke went north and saw Lake Victoria. He thought that this huge lake must be the source of the Nile. But he could not prove it.

Speke set out again in 1860 with an officer, James Grant. In 1862, they discovered the main outlet of Lake Victoria. This is, in fact, one of the main sources of the White Nile. Speke and Grant were too tired to trace the river northwards and so people still did not know whether it was the river's source. This is why explorers continued the search for the source of the White Nile for many years. In doing so, they helped to fill in the map of Africa.

Villages on the River Zaire often contained hostile Africans who attacked Stanley's party. Fighting and disease caused the deaths of many of Stanley's men.

Stanley was a bold and sometimes ruthless man. If his porters threatened to run away or lose his equipment in the river, he would frighten them with his pistol.

Stanley's boat, the *Lady Alice*, could be divided into five pieces so that it could be carried when Stanley was marching overland or skirting waterfalls and rapids.

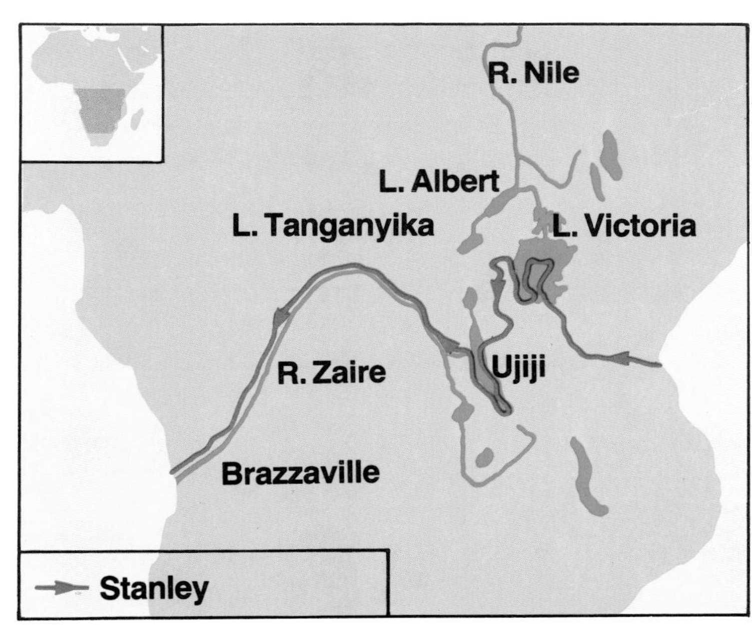

The map (above) shows the route taken by Stanley on his great journey of exploration across Africa between November 1874 and August 1877.

Africans in canoes (right) attacked Stanley when he was exploring Lake Victoria. Stanley returned later with soldiers to take revenge on his attackers.

The explorer Henry Morton Stanley was born in Wales. He was originally named John Rowlands but, at the age of 18, he emigrated to the United States. There, an American merchant adopted him and gave him his own name. Stanley served in the American Civil War (1861-5) and, in 1867, he joined the *New York Herald* as a reporter.

While working for this newspaper, he accepted a mission to go to Africa to look for David Livingstone. No news had come from the explorer for some time. On his way to Africa, Stanley saw the opening of the Suez Canal in Egypt in 1869. He went on to India and, finally, he reached Zanzibar in January 1870. He set out in March with 192 men. After suffering many hardships, Stanley found Livingstone in October 1871 (see page 27). The two men became good friends. They spent five months together and explored the northern end of Lake Tanganyika. They found that a river there, called the Ruzizi, flowed *into* the lake, not out of it. This meant that it was not the source of the Nile, for which Livingstone had been searching. Stanley was fascinated by Livingstone's work and greatly admired him. He described their meeting in his book *How I Found Livingstone* (1872).

Stanley left Livingstone in March 1872 and returned to London. He could not forget Livingstone and, after he heard of the explorer's death, he planned a journey to follow up Livingstone's work. Stanley wanted to explore lakes Victoria and Tanganyika and travel down the River Lualaba, west of Lake Tanganyika, because Livingstone had thought that this river might flow into the Nile.

In November 1874, Stanley marched into the interior of Africa. His party was attacked and 26 men were killed. Disease also struck and, in three months, a quarter of his party of 356 were dead. He reached Lake Victoria in February 1875. He explored the lake and proved it was not several small lakes as some people thought.

Stanley wanted to go on to Lake Albert. The Nile flows from Lake Victoria into the northern end of Lake Albert and out again. Another explorer, Sir Samuel Baker, had seen this in 1864. Lake Albert (now called Mobutu Sese Seko) is, therefore, part of the Nile system. Wars in the area stopped Stanley going to Lake Albert. Instead, he went south to Lake Tanganyika and sailed around it.

The party then turned to the River Lualaba. This river flowed north from Lake Tanganyika through thick forests. Cannibals, river rapids and disease all lay ahead. On Boxing Day 1876, Stanley's party contained 149 men, women and children. Only 114 people were left when the party reached the Atlantic. The boat, *Lady Alice,* which the party had used was abandoned when Stanley was within a few days march of the coast. Stanley had proved that the River Lualaba was not the Nile. It was part of the River Congo (now called the Zaire).

In 1887, Stanley made another journey to Africa. He went to search for the German governor of the southern Sudan, who had been cut off during a war in Sudan between Britain and local Muslims. This man was known by a name he had adopted, Emin Pasha. Stanley found Emin Pasha and rescued him. He also discovered Lake Edward on this trip and found that the River Semliki, which drains it, flows into Lake Albert. Stanley's other discovery was the lofty Ruwenzori mountains, which are also called the Mountains of the Moon.

By 1890, Stanley was back in London. He gave up exploring and became a British citizen once more. He became a member of parliament in 1895 and, in 1899, he was made a knight.

Exploring the South Pacific

People once thought that men with no heads lived on New Guinea. Explorers proved that most stories of monsters in far-away lands were untrue.

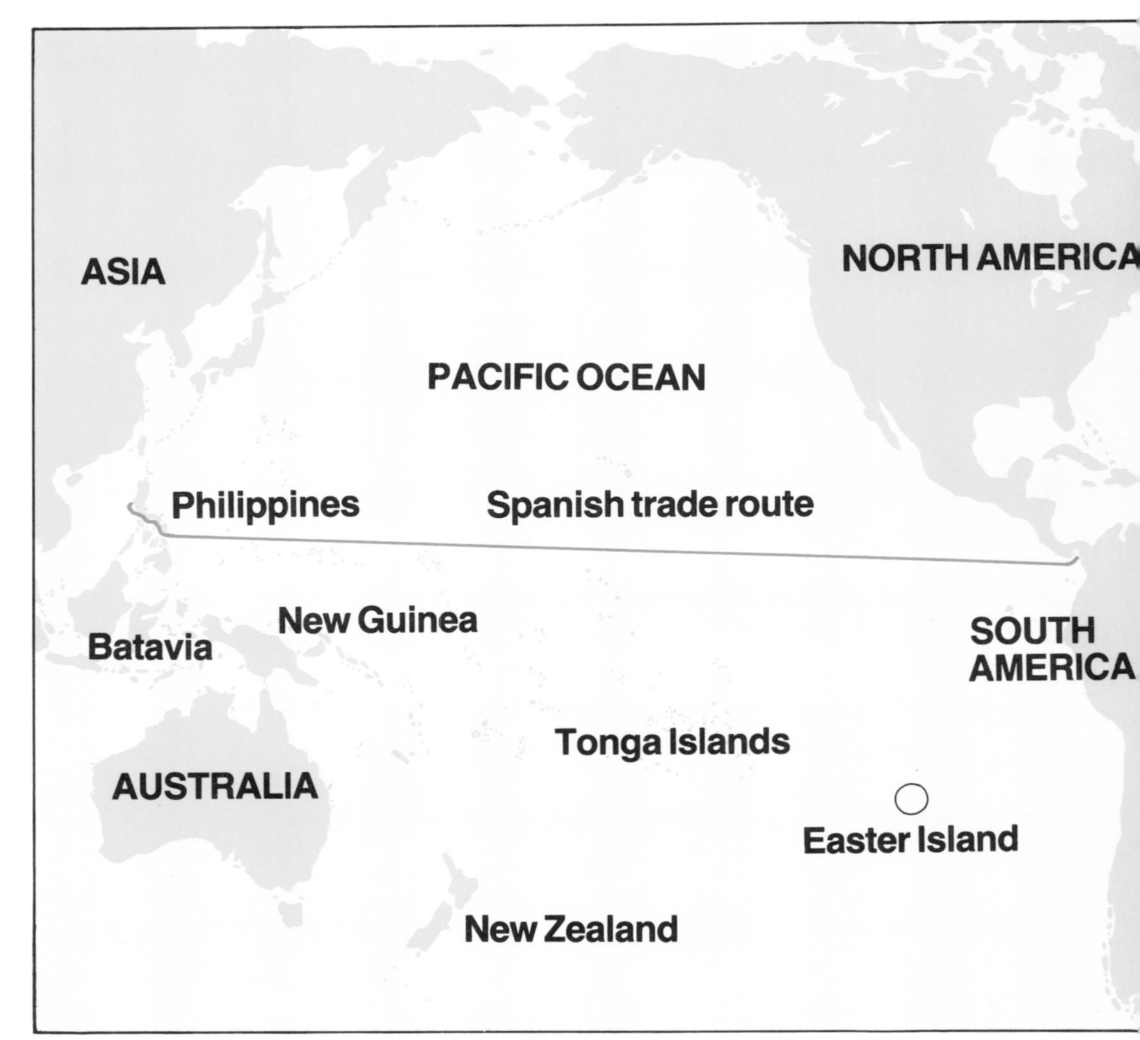

The Pacific Ocean is the world's largest ocean. For hundreds of years before Europeans saw the ocean, the Polynesians travelled around it in dugout canoes. They found their way from the positions of the Sun and the stars in the sky. The Portuguese discovered the Spice Islands (now the Moluccas) in 1511 and they may have seen Australia. Spanish explorers sailed from Central and South America. In the 17th century the Dutch explored the Pacific. Later, other countries did too.

The Polynesians, who live on islands in the Pacific Ocean, are skilled sailors. Polynesia contains the islands found within a triangle formed by New Zealand, Hawaii and Easter Island.

The French explorer Louis Antoine de Bougainville sailed around the world between 1766 and 1769. He came close to finding eastern Australia but, near Australia, he turned north to avoid dangerous rocks and reefs.

Large war canoes were used by the people of Tahiti, a large island in the South Pacific Ocean. This picture was painted by William Hodges, an artist who sailed with Captain James Cook. The picture shows the islanders preparing to invade another island in 1774. Cook recorded that the whole war fleet contained hundreds of canoes. The Pacific islanders sometimes attacked European explorers.

Easter Island contains many huge stone statues. They were first seen by a Dutchman, Jacob Roggeveen, on Easter Sunday, 1772.

Maps of the 16th century show a large continent in the southern hemisphere. This continent was usually drawn so that it stretched from the South Pole to the equator. No one had seen this continent. But people, since the ancient Greeks, believed that there must be a continent in the south which balanced North America, Europe and Asia in the northern hemisphere. They called the southern continent *Terra Australis Incognita.* This means the 'unknown southern land'.

The first Europeans to see the southern continent of Australia were probably the Portuguese. They had discovered the Spice Islands (now the Moluccas) in 1511 and they soon controlled trade in the East Indies. They probably found northern Australia, but they kept their discovery a secret, thinking perhaps that knowledge of Australia's existence might help their rivals.

Spain also had interests in the Pacific Ocean. The Philippines became their main trading centre after the voyage of Ferdinand Magellan (see page 12). However, both Magellan and Sir Francis Drake sailed north of Australia. Other Spanish sailors went from Peru and Mexico to explore the Pacific, but they did not find any southern continent.

Spanish exploration in the Pacific ended in the early 1600s. This was because Spain's power had declined. In 1588, the English navy had destroyed the Spanish armada (a fleet of warships). Earlier, in 1568, the Netherlands had revolted against Spanish rule and the Dutch had declared their nation independent in 1581.

The Dutch took control of the Pacific region from Spain. They drove out the Portuguese from the Spice Islands and Sri Lanka. Their trade in the East Indies was carried on by the Dutch East India Company. The centre of this powerful company was at Batavia (now Jakarta) on the island of Java. The Dutch were more interested in trade and profit than in exploration. However, it was a Dutchman who first saw Australia.

The Dutch founded Batavia (now Jakarta, in Indonesia) in 1619 as their trading centre in the East Indies.

The Dutch carried out important exploratory work in the Pacific in the 17th century. They were not interested in finding out about the Pacific. They explored to increase trade and to become wealthy.

The first sight of Australia, by a European, about which we are certain, was by a Dutchman called Willem Jansz.He sighted Cape York, the peninsula in north-eastern Australia. He then explored the coast of the Gulf of Carpentaria. In 1616, another Dutchman, Dirk Hartog, became the first European to land in western Australia.

The most important Dutch expedition, however, was inspired by the governor of the Dutch East India Company, Anthony van Diemen. It was led by Abel Janszoon Tasman. Tasman sailed from Batavia in August 1642 to the island of Mauritius in the Indian Ocean. He then turned east but he saw nothing until he reached a large island on November 24. He called this island Van Diemen's Land. (We now call it Tasmania. It was renamed in 1856.) Tasman sailed on and discovered South Island, New Zealand. He finally got back to Batavia in June 1643. He had sailed all the way around Australia, but he had not seen the huge continent at all during his voyage.

The Dutch called Australia 'New Holland', but it was not like their homeland. The land was barren and the people were poor and often war-like. The place seemed to have no value and the Dutch lost interest.

At the end of the 17th century, Britain sent an expedition to explore the South Seas. William Dampier was the first Englishman to sight Australia. Dampier, a former pirate but also a scholarly man, set out in 1699. He explored the dry coast of north-western Australia before going on to the East Indies. He agreed that Australia was a hostile land. This view was held until Captain Cook discovered the eastern coast.

The Dutchman Abel Tasman sailed around Australia in 1642-3. He discovered Tasmania and New Zealand, but not Australia.

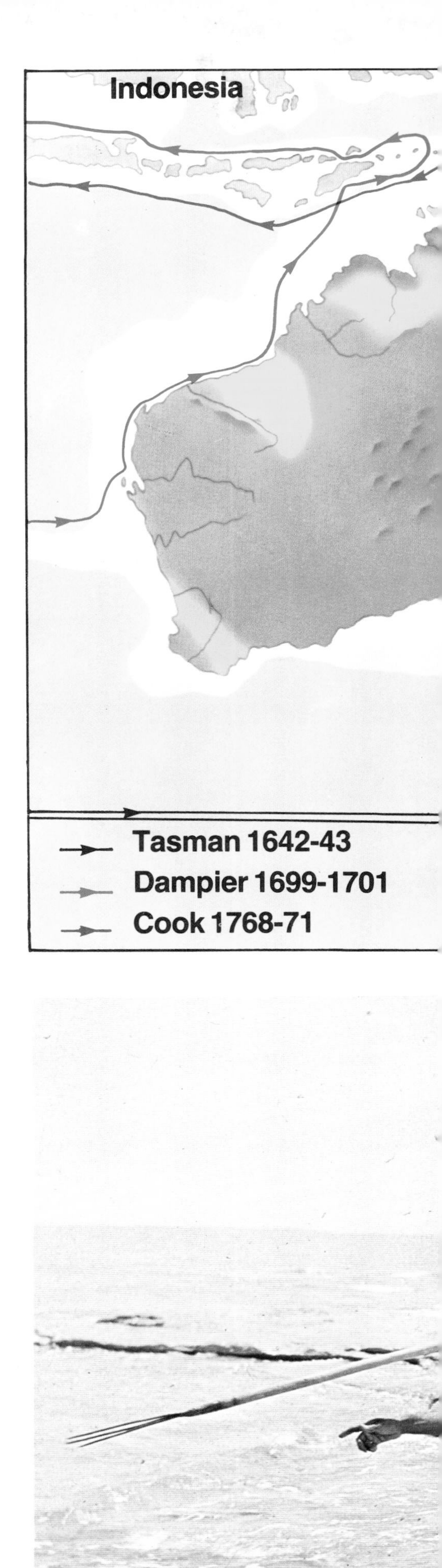

This Australian Aborigine is trying to harpoon a fish. The Aborigines probably reached Australia from Asia about 20,000 years ago.

Australia has many animals which are not found elsewhere. This is because Australia has been surrounded by the oceans for about 50 million years and animal life developed independently. This kangaroo was painted by George Stubbs. It was based on a skin brought back to Britain by Captain Cook.

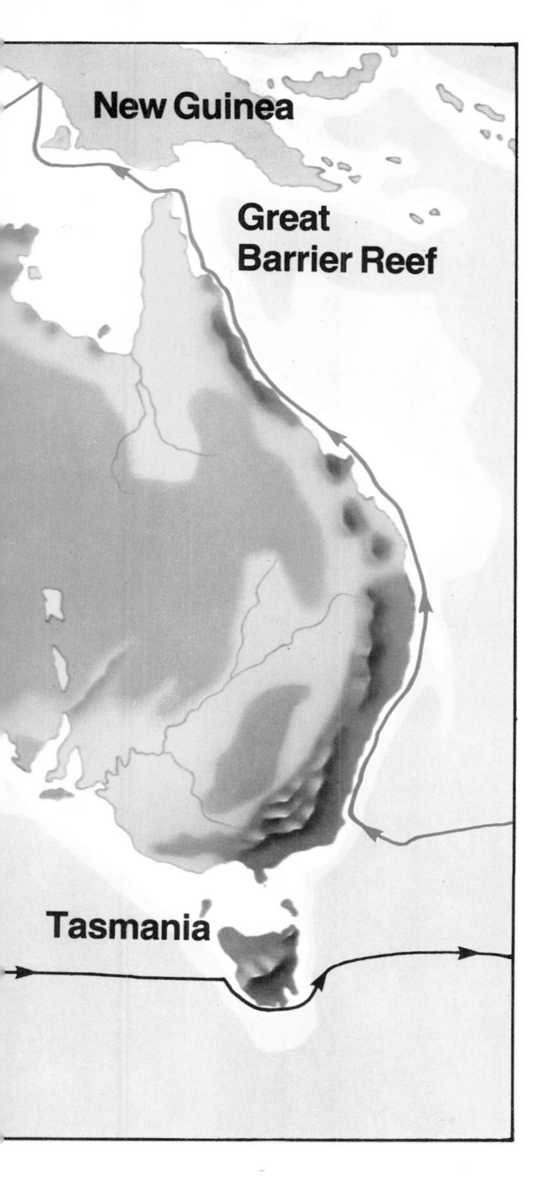

The map (left) shows three important voyages of discovery. They all added to people's knowledge of Australia.

William Dampier (above), a former pirate, was the first Englishman to set foot in Australia.

The Aborigines hunted animals and gathered plants. They were nomads who moved around in search of food and water. Some Aborigines threatened early explorers, so they said that Australia was a hostile place.

Before photography was invented, artists often accompanied explorers. This Aborigine was sketched by an artist who sailed with a French explorer, Nicolas Baudin, on a voyage of exploration in southern Australia in 1800-2.

This carving of a war god was given to Captain Cook in Hawaii. When Cook first arrived in Hawaii in 1779, the people thought that he was a god.

This is a copy of the first successful chronometer, a very accurate clock made in about 1760. Cook used a chronometer to find his longitude. He measured the local time from the position of the Sun. He compared it with the time at Greenwich (0° longitude) shown on the chronometer. He worked out the longitude from the time difference.

The painting shows the Maoris of New Zealand performing a *haka*, or war dance, before a battle.

A mother relaxes with her children (above) in Fiji. Cook visited Fiji in 1774 on his second great voyage.

Cook (right) was killed in Hawaii on 14 February 1779. He was trying to order his men to stop shooting.

James Cook, the son of a poor farm worker in Yorkshire, joined the British navy as an able seaman in 1755. He learned much about astronomy and map-making and he gradually rose in rank. In 1768, Cook was appointed commander of an expedition to explore the Pacific Ocean.

Cook left Plymouth in August, 1768. He took several scientists with him. One of them was a young naturalist, Joseph Banks, who later sponsored explorers in Africa (see page 24). Cook reached Tahiti in 1769, where he carried out observations of the planet Venus.

In July, he set out for New Zealand, where he mapped the coasts. He managed to prove that New Zealand was not part of a southern continent as people had believed. Cook met many Maoris in New Zealand. He liked them, but he did not approve of the fact that some were cannibals.

After mapping New Zealand, Cook sailed west. In April 1770, he sighted the south-east corner of Australia. He turned north and sailed along the coast. He claimed this pleasant part of eastern Australia for Britain. It eventually became a penal settlement.

In 1772, Cook set out on his second voyage. He went south and sailed beyond the Antarctic Circle. He went on to New Zealand before sailing south again. He got within about 2,000 kilometres of the South Pole when his way was blocked by ice. He returned and mapped many Pacific islands before going home.

In 1776, he began his third voyage. He discovered the Hawaiian islands in 1777. In 1778 he sailed on to explore the coasts of Canada and Alaska and the Bering Strait. Cook was looking for a North-West Passage (see pages 38-39). But he turned back when he saw a vast sea of ice in the freezing Arctic Ocean.

The Maoris fought back Cook's men who tried to land after arriving in New Zealand in 1769.

James Cook (above) was born in Yorkshire. He was the greatest of all the British sailor-explorers. He sailed every ocean and made three great voyages of discovery in the Pacific Ocean.

The map (below) shows parts of Captain Cook's three voyages in the Pacific Ocean. Cook proved that there was no large continent like Eurasia in the southern continent.

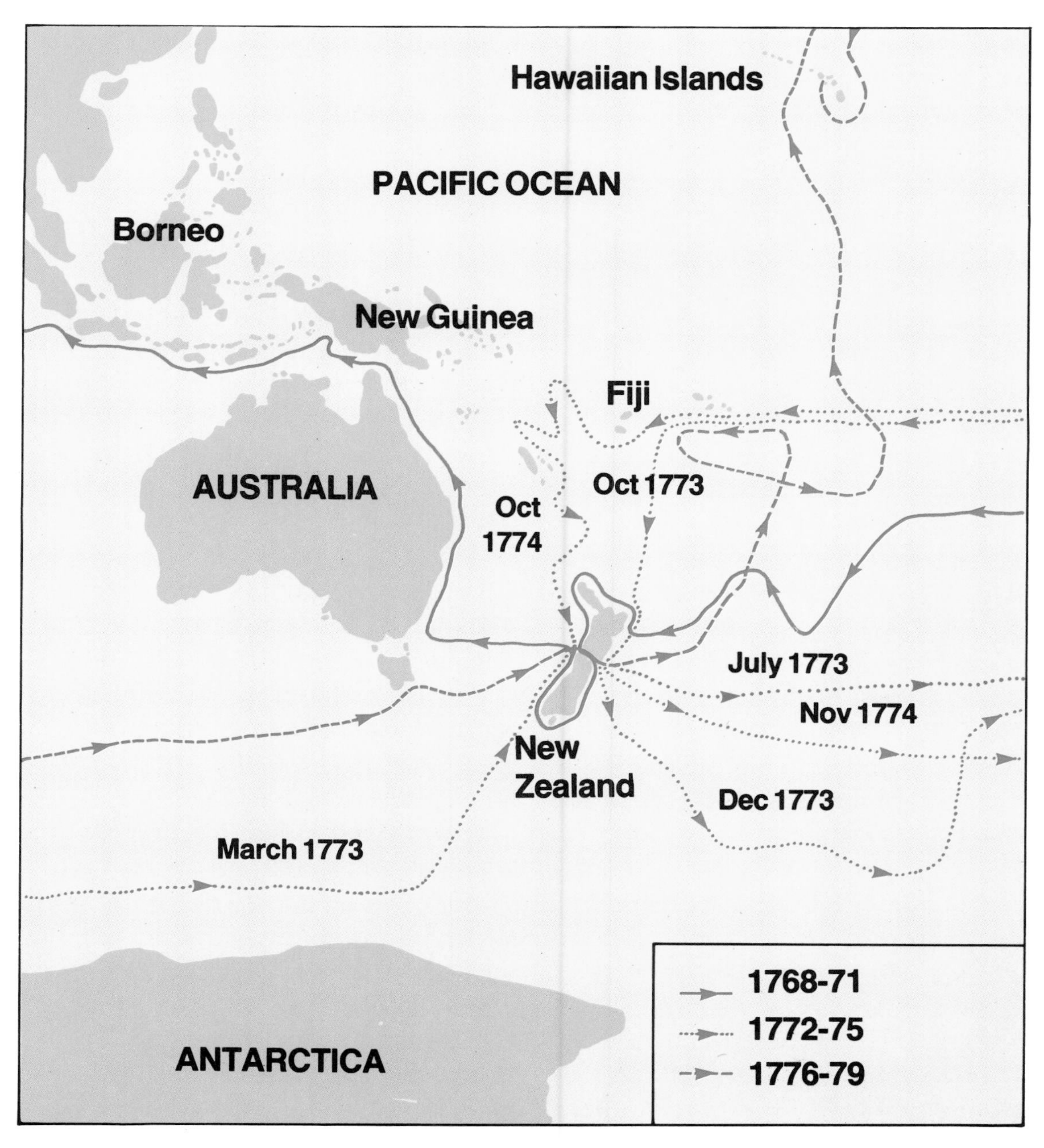

Cook went back to Hawaii. He was treated at first like a god but, on 13 February 1779, one of his ship's boats was stolen. The next day Cook went ashore and was killed in a battle with the local tribesmen over the boat which had been stolen.

James Cook was the greatest of all the British sailor-explorers. He had proved that there was no southern continent stretching from the South Pole northwards into the Pacific. He had mapped New Zealand, discovered the fertile coasts of eastern Australia and filled in most of the gaps on the map of the Pacific Ocean.

Following Cook's exploration of the east coast of Australia, a settlement for British convicts was established in New South Wales in January 1788.

The picture shows the convicts building their new homes on the site of what is now Sydney. The first free settlers entered Australia in 1793.

Eastern Australia was proved to be part of the same continent as New Holland by Matthew Flinders. He sailed right around Australia. Flinders explored south-eastern Australia and Tasmania with another navigator, George Bass, in 1798-9. In 1800, Flinders went to Britain to win support for his plan to map Australia's coasts. He set out in July 1801 and reached the south-west corner of Australia in December. He sailed along the southern coast and on to Sydney. He continued north and began to map the coast of the Gulf of Carpentaria when he found that his ship was leaking badly. He was forced to speed up his voyage and he arrived at Sydney again in July 1803.

Exploration of the interior took much longer. High mountains near the eastern coast made inland travel difficult. The mountains behind Sydney were first crossed in 1813 and fertile land was found on the other side. In 1824, Hamilton Hume, the first Australian-born explorer, travelled through the south-east. He reached the coast near where Melbourne stands today. The rivers of the south-east are long and large. Some people thought that they came from an inland sea. This idea was disproved by Charles Sturt who explored them between 1828 and 1830.

The dry central and western part of Australia remained unexplored. Great heat, lack of water and hostile Aborigines made it dangerous.

Charles Sturt tried to reach the centre of Australia in 1844-5. He got within 240 kilometres but he had to turn back as he was going blind. At the same time, Ludwig Leichhardt was exploring northern Australia. In 1848, Leichhardt set out to cross central Australia. But his party disappeared. They were probably killed by Aborigines.

In 1860, South Australia's government offered £2,000 to the first man to cross Australia from south to north. John McDouall Stuart set out in April 1860. He reached the north coast in July 1862. But he had been beaten by another expedition which had ended in tragedy.

Robert O'Hara Burke and William Wills had set out from Melbourne in August 1860 on the same mission. They set up a depot at Cooper's Creek. In December, Burke and Wills set out with Charles Gray and John King. Slowed down by swampy land, they reached the sea in February 1861. They were short of supplies. On their way back, Gray collapsed and died. The others reached their depot but found that the other men and the supplies had gone. Burke and Wills died of starvation in June. King was luckier. He was found by Aborigines who helped him. He was rescued three months later.

Ludwig Leichhardt was a German-born explorer who went to Australia in 1842. He explored the eastern coastlands and the northern interior in 1844 and 1845. He set up an expedition to find a practical route from Moreton Bay (now Brisbane) to Perth. He set out in 1848 but he and his party of seven men completely disappeared. No one knows what happened to them.

Robert O'Hara Burke and William Wills led the first successful overland expedition across Australia from north to south in 1860-1. Camels carried their provisions. Both men died on the return journey.

Charles Sturt explored parts of Australia between the 1820s and the 1840s.

John McDouall Stuart, a Scot, crossed Australia from south to north in 1860-2.

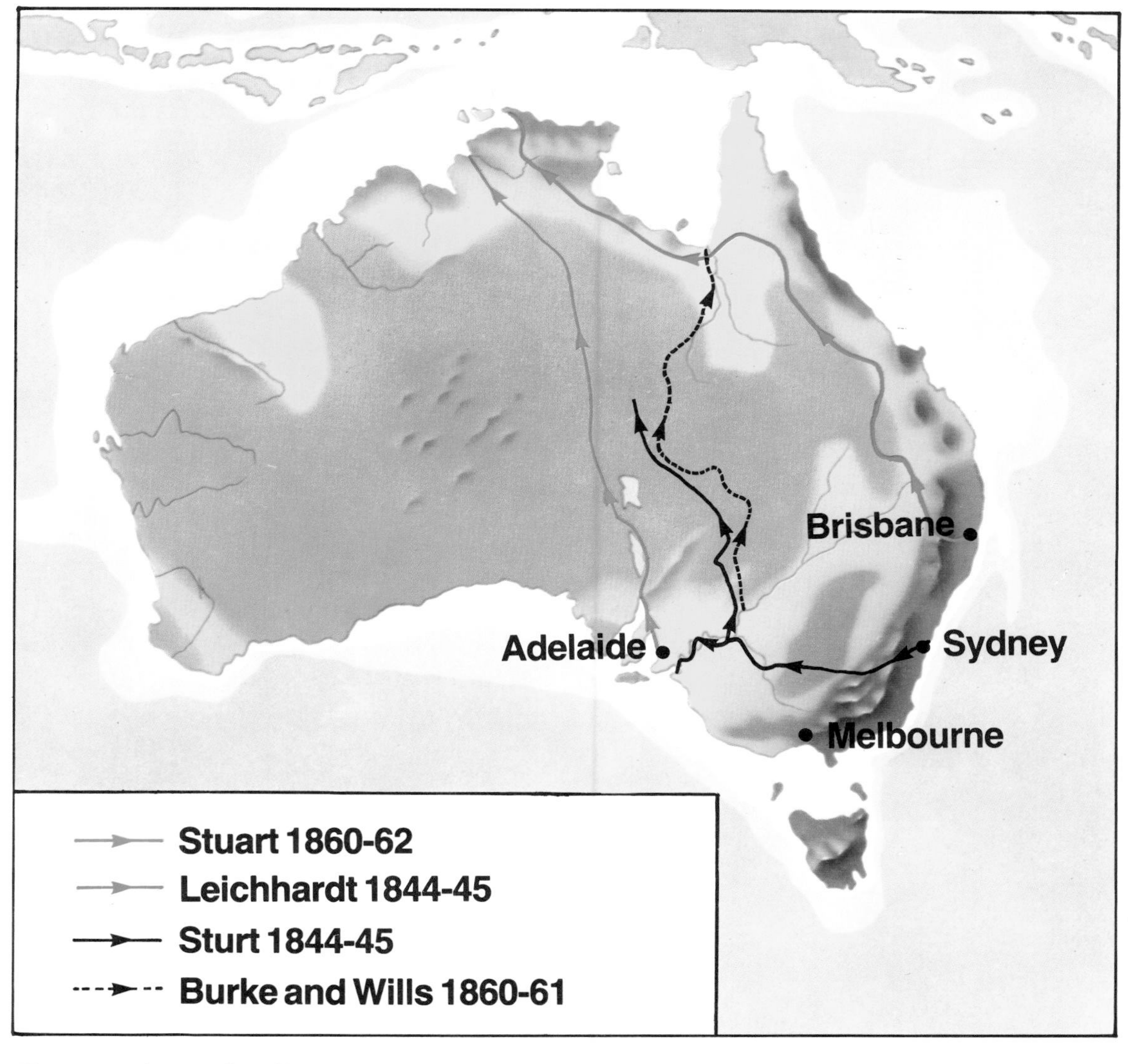

The map shows that Australia's four largest cities – Brisbane, Sydney, Melbourne and Adelaide – are all in the fertile south-east, where the first settlers went. The map also shows some of the routes taken by the brave explorers who ventured into the hot deserts of the interior.

The North-West Passage

The search for a North-West Passage to Asia inspired many explorers. Many men tried to find the route but failed. Some died in the attempt. But these explorers often made other important discoveries. For example, French explorers searched for the North-West Passage in the 16th and 17th centuries. They failed but they explored the St Lawrence River and the Great Lakes regions of North America (see pages 16-17).

England also took part in the search. In 1576, Sir Martin Frobisher sailed to Greenland and on to Baffin Island. There he saw Eskimos. He took one back with him. He believed that Eskimos were Asians because they resembled the Chinese.

The most important early explorer was Henry Hudson. On his last voyage in 1610, Hudson sailed into Hudson Bay. He thought that this was the way to Asia. But he died in 1611 after his crew cast him adrift. In 1616, another explorer, William Baffin, found a waterway, called Lancaster Sound, between two Arctic islands. He sailed into it, but was forced back by thick sea ice. Had he gone on, he would have been the first man to discover the North-West Passage.

The Arctic parts of northern Canada are extremely cold and temperatures drop to −46°C. But explorers still sought a way to Asia. Activity was great in the early 19th century, when such men as John Ross, James Clark Ross and Edward Parry made important discoveries in Arctic waters. In 1845, the explorer, Sir John Franklin, set out with 129 men. They were never seen again. Many search parties hunted for them. No one was saved but the rescue parties filled in many gaps on the map.

In 1903, Roald Amundsen made a voyage to North America to do scientific research. On this trip, which ended in 1906, he made the first successful voyage through the North-West Passage. Recently, prospectors have found rich deposits of oil in northern Alaska and valuable metals in northern Canada. To get these products out, the North-West Passage might still have to be used.

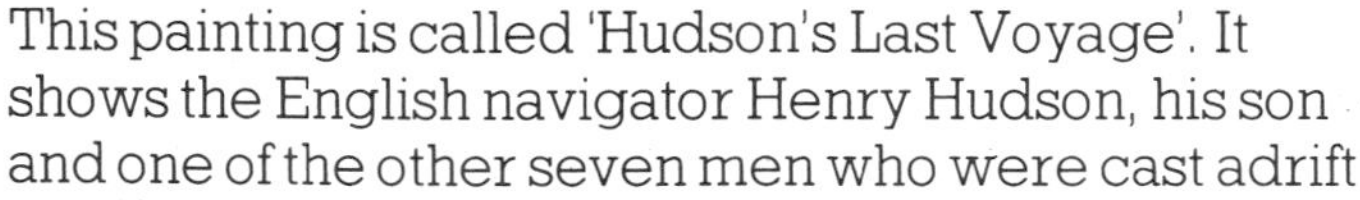

This painting is called 'Hudson's Last Voyage'. It shows the English navigator Henry Hudson, his son and one of the other seven men who were cast adrift in James Bay, part of Hudson Bay, in North America, in 1611. They were never seen again. They had been searching for a North-West Passage to Asia.

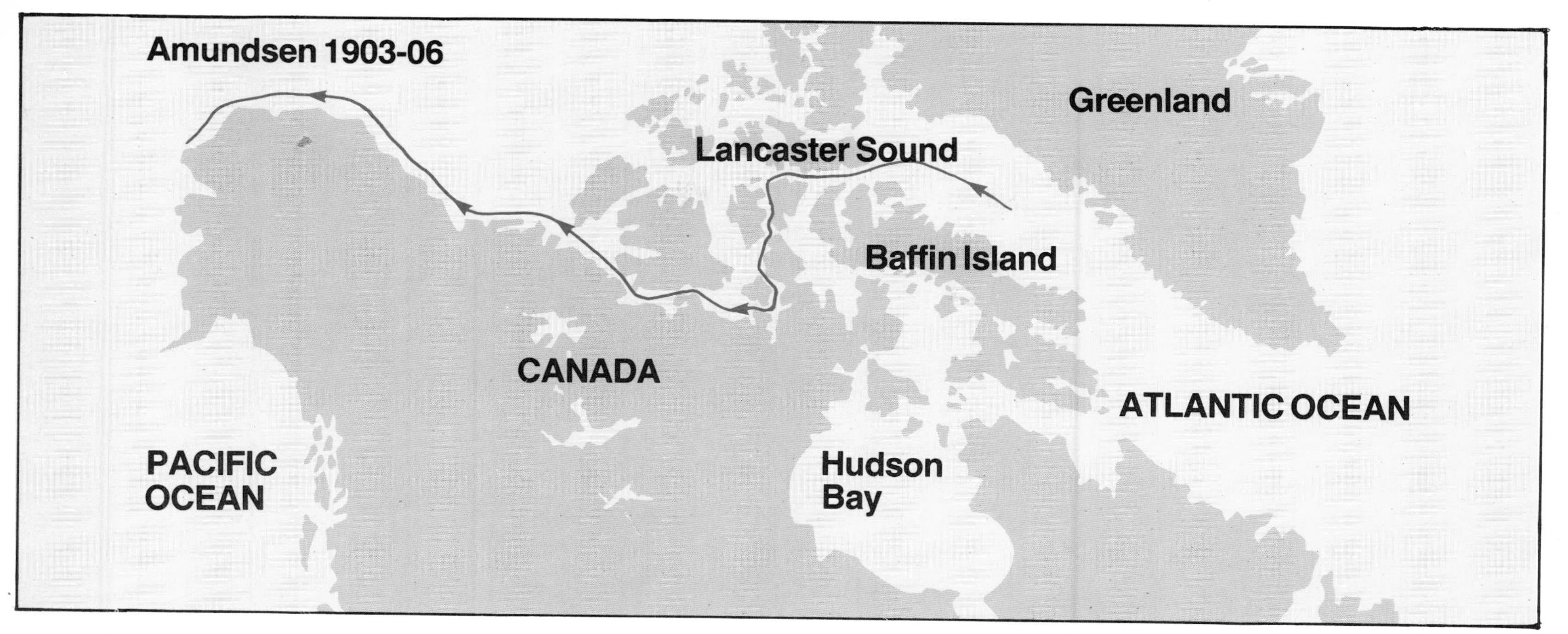

The map shows the route taken by the Norwegian explorer Roald Amundsen. He successfully completed a voyage through the North-West Passage from the Atlantic Ocean to the Pacific Ocean between 1903 and 1906. This route had been sought since the days of Columbus. It is not an important route now.

The *Fox* was bought by Lady Franklin. It was sent to the Arctic in the 1850s to look for her missing husband. He had disappeared while searching for the North-West Passage.

This Eskimo was painted in about 1830. An Eskimo was taken to England in 1576 but died of a cold.

This oil tanker went through the North-West Passage in 1969 to find if oil could be exported this way.

The *Gjöa* was the ship used by Roald Amundsen when he passed through the North-West Passage.

The North-East Passage

The desire to find a route to Asia led Vasco da Gama around Africa (pages 22-23) and Ferdinand Magellan around the stormy tip of South America (pages 12-13). Many other explorers looked for a North-West Passage (pages 38-39). There was one other route to Asia for Europeans to explore. This was the route through the Arctic Ocean around northern Europe and Asia. It was called the North-East Passage.

In the late 16th century, English merchants were envious of the rich Portuguese and Spanish merchants who traded in Asia. In 1553, an English company of merchant adventurers sponsored an expedition to find a North-East Passage. It was led by Sir Hugh Willoughby and Richard Chancellor. The ships ran into a storm off Norway and Willoughby was forced ashore where he died. Chancellor went on. He reached the mouth of the Dvina River in northern Russia. He met officials of the Tsar Ivan the Terrible and then went home. His voyage led to the founding of the Muscovy Company, which traded profitably with Russia.

The Dutch also became interested in the North-East Passage. In the 1590s, a Dutch explorer, Willem Barents, made three voyages into the Arctic Ocean. He discovered Spitzbergen (now Svalbard) in 1596. But his ship was damaged by ice. He died on the long journey home.

In the 1720s, the Russian emperor Peter the Great wanted to know more about the Pacific coast of his vast empire. He sent a Danish explorer, Vitus Bering, to the Pacific coast to explore the Kamchatka Peninsula. In 1728, Bering sailed north along the peninsula. He went through the strait which now bears his name. The strait separates Russia from Alaska and it also joins the Pacific to the Arctic Ocean. Bering explored some of the Arctic coast, but bad weather made him turn back. In 1741, Bering returned to the strait and sighted North America.

However, the first successful voyage through the icy North-East Passage did not take place until 1878-9. This feat was achieved by the Swedish explorer Nils Nordenskjöld after a long and difficult voyage. Today the North-East Passage is an important route for Russian and other ships. But ice-breakers must be used to smash up the ice.

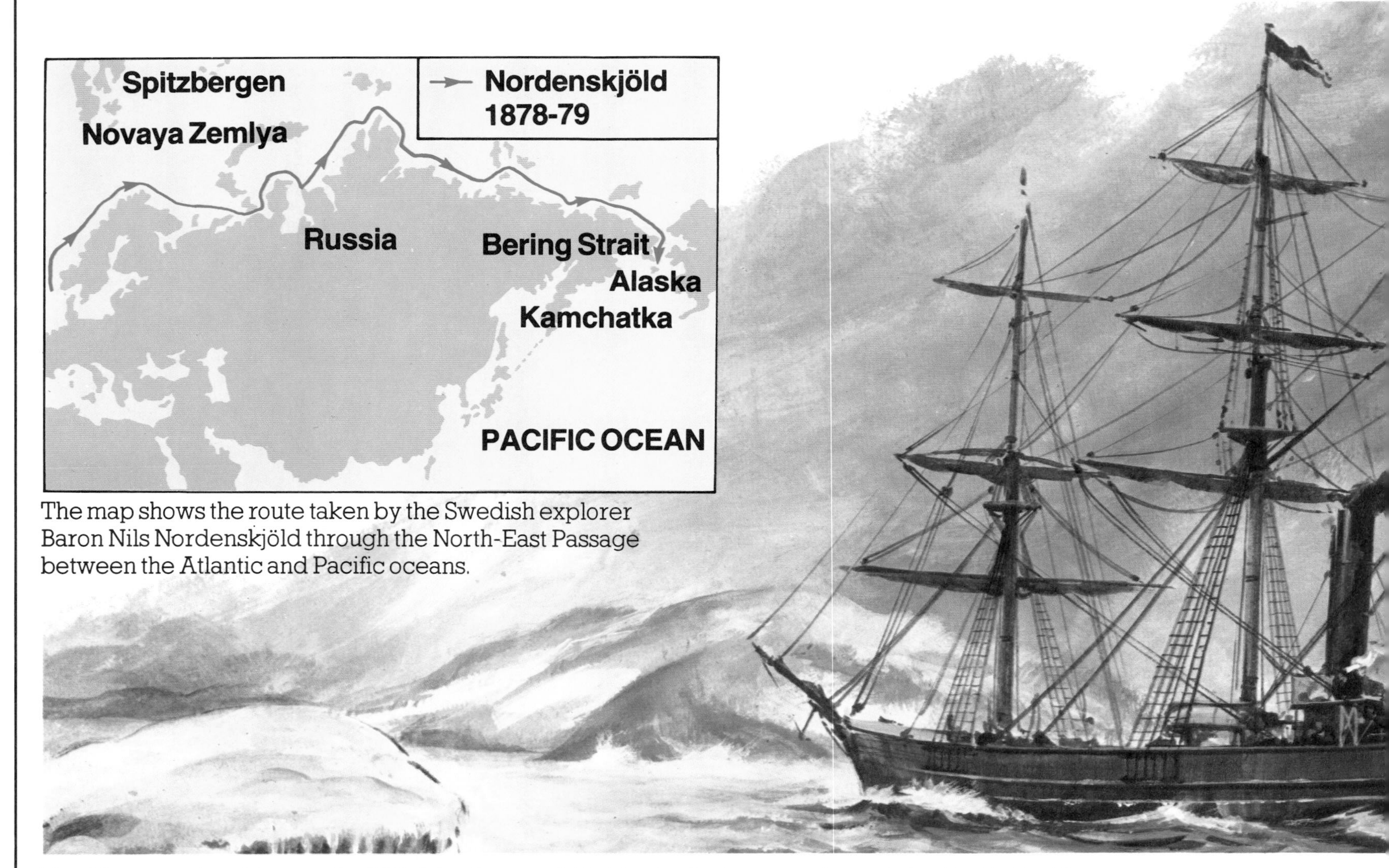

The map shows the route taken by the Swedish explorer Baron Nils Nordenskjöld through the North-East Passage between the Atlantic and Pacific oceans.

The *Vega* was the ship used in 1878-9 by Nordenskjöld for the first voyage through the North-East Passage.

Vitus Bering, a Dane, served the Russian Tsar, Peter the Great. Bering died in 1741 on an island named after him, following his exploration of the Bering Strait. This strait separates North America from Asia.

This painting shows the hut built by Willem Barents, a Dutch explorer, and his men on Novaya Zemlya, a Russian island in the Arctic. Barents' men had been forced to winter on this island, because their ship had been damaged by the ice, while they were searching for the North-East Passage. The picture shows an event which occurred on 12 February 1597. This was the day when the men shot a polar bear, which is on the left of the picture. Barents survived most of the winter, but died on the way home from this expedition.

Fridtjof Nansen, a Norwegian explorer, led an expedition in 1893-6. This took him closer to the North Pole than any other man before him.

The *Fram* was specially designed by Fridtjof Nansen to withstand the great pressure exerted by the sea ice in the Arctic Ocean. Nansen planned to get the *Fram* jammed in the ice near the New Siberian Islands. He believed that the ship would drift from east to west, carried by an ocean current. The *Fram* drifted near the North Pole.

Robert Peary, a United States naval officer, reached the North Pole on 6 April 1909. He was accompanied by his black servant, Matthew Henson, and four Eskimos.

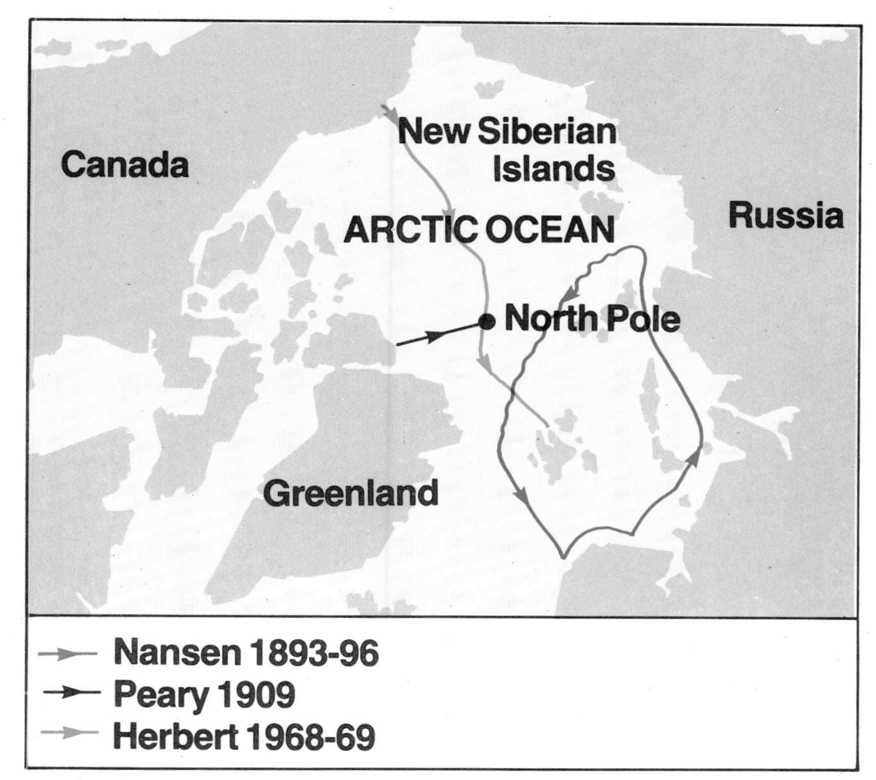

The map (above) shows the routes taken by Fridtjof Nansen in the *Fram*; by Peary when he reached the North Pole; and by Wally Herbert who crossed the icy ocean on foot.

The Italian, Umberto Nobile (left), flew over the North Pole in 1926. He did the same thing in 1928 in the *Italia*, but this airship crashed on the way back.

In 1895, Nansen and Hjalmar Johansen tried to reach the North Pole with dog sledges.

In the late 19th century, a race began between various explorers to reach the North Pole. One American expedition in 1881 ended when a ship, the *Jeannette,* was crushed in the ice near the New Siberian Islands. Equipment from the ship was found on south-western Greenland three years later. The Norwegian, Fridtjof Nansen, believed that an ocean current swept across the North Pole and had carried these items with it. Nansen decided the current could carry a ship to the North Pole.

Nansen designed a ship, the *Fram,* to withstand pressure from the ice. He sailed the *Fram* into the Arctic. By September 1893, it was locked in the ice near the New Siberian Islands. Slowly, it drifted northwards. When it reached about latitude 84° North, it started to drift west. Nansen and one of his officers left the ship and in April 1895, on foot, they got nearer the Pole than anyone before them. They turned back, but the *Fram* went on drifting. It finally broke out from the ice at Spitzbergen in 1896.

The man who first reached the North Pole was an American naval officer, Robert Peary. Peary had made several previous expeditions to Greenland and the Arctic. He reached the Pole on 6 April 1909. In 1926, Richard Byrd and Floyd Bennett became the first men to fly over the North Pole. In the same year, Roald Amundsen and an American, Lincoln Ellsworth, joined the Italian pilot Umberto Nobile in his airship *Norge* to fly from Spitzbergen, over the North Pole and on to Alaska.

The first surface crossing of the Arctic, from Alaska to Spitzbergen via the North Pole, was made by an expedition led by a Briton, Wally Herbert, in 1968-69. A distance of 6,000 kilometres was covered in 476 days.

Captain Cook explored the southern oceans in the 1700s. He sailed all round the continent of Antartica, but he did not see land. In the early 1800s, seal and whale hunters sailed farther and farther south. They were probably the first men to see and set foot on Antarctica.

In the 1840s, parts of the coast of Antarctica were explored by expeditions searching for the South Magnetic Pole. French and American ships sighted and explored areas of the coast in 1840. A British expedition made some important discoveries in 1839-42. The expedition was led by the Arctic explorer, James Clark Ross. Ross discovered the Ross Ice Shelf, bordering the Ross Sea. This became the starting-point for later explorers who braved the cold to reach the South Pole.

In June 1910, Robert Falcon Scott set out from Britain to conquer the South Pole. But he had a rival. The Norwegian, Roald Amundsen, had also set his sights on the Pole.

Amundsen went to Antarctica in Nansen's ship, the *Fram*. On 11 October 1911, he set out over the Ross Ice Shelf with four companions, some light-weight sledges and 52 huskies. He reached the Pole on December 14. His journey of 3,000 km had taken 99 days.

Scott set out on November 1 by a different route. He had 100 km farther to go than Amundsen. He ran into bad weather then his ponies sank in the deep snow and had to be shot. The explorers then had to pull the sledges themselves. They reached the South Pole on 17 January 1912. They were disappointed to find that Amundsen had arrived before them. On the return journey, Edgar Evans was injured and he died. Captain Oates, who was ill, walked out into a blizzard and never returned. Finally, the three survivors were trapped in their tent by a fierce blizzard. They died there, 10 km from their food depot.

Since 1912, explorers and scientists have continued to work in Antartica, the last continent to be discovered.

This Sno-Cat is a vehicle designed for travel in Antarctica. But it nearly vanished into crevasses more than once during the Commonwealth Trans-Antarctic Expedition in 1957 and 1958. The expedition was commanded by Sir Vivian Fuchs.

Scott's party reached the South Pole on 17 January 1912. They raised the Union Jack near the Norwegian flag. This had been left by Roald Amundsen, who, unknown to Scott's party, had reached the Pole on 14 December 1911. From left to right, are Captain Lawrence Oates, Lieutenant Henry Bowers (seated), Captain Robert Falcon Scott, Dr Edward Wilson (also seated) and Petty Officer Edgar Evans.

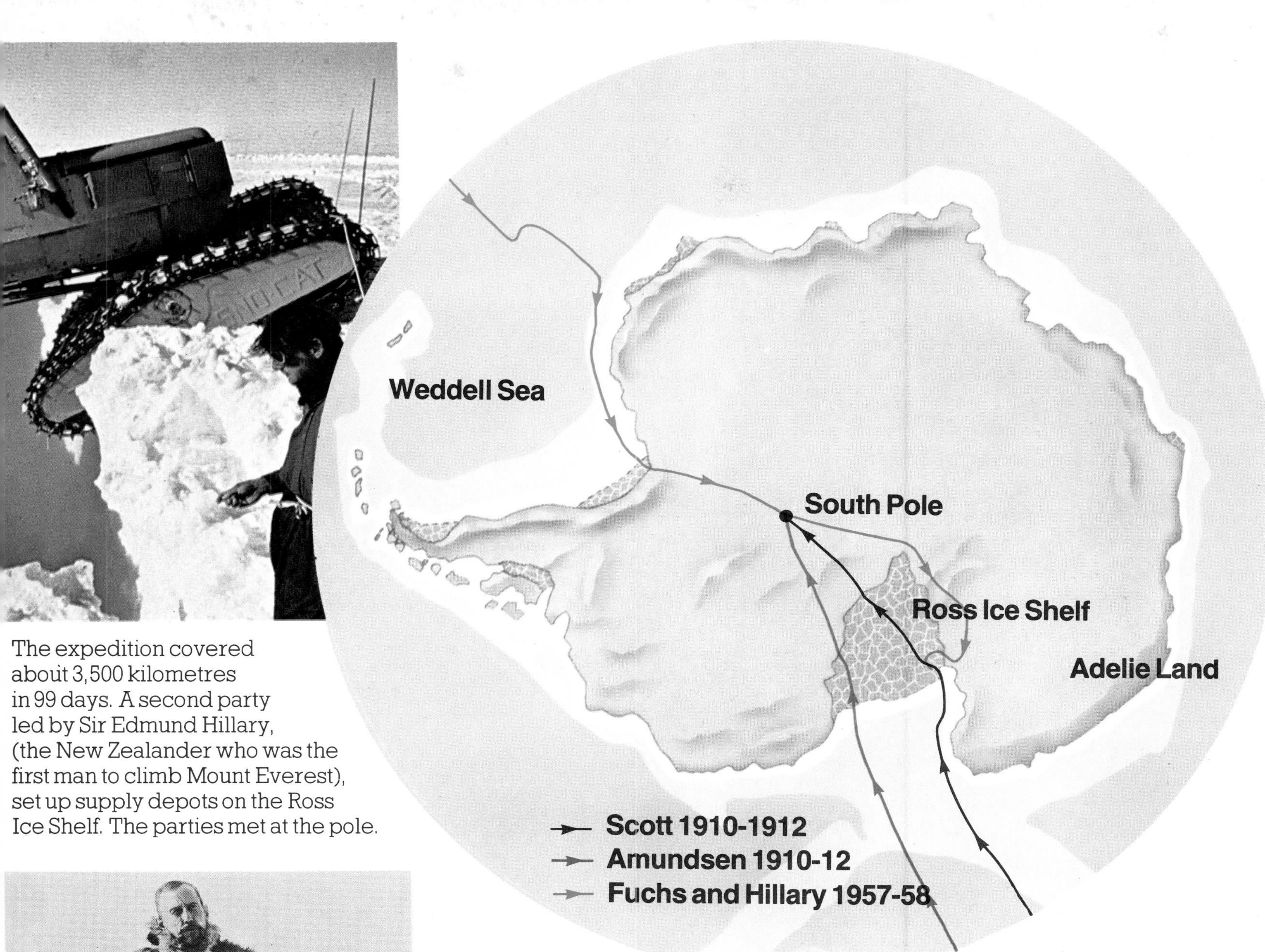

The expedition covered about 3,500 kilometres in 99 days. A second party led by Sir Edmund Hillary, (the New Zealander who was the first man to climb Mount Everest), set up supply depots on the Ross Ice Shelf. The parties met at the pole.

The map shows three great journeys of exploration in Antarctica, the coldest place on Earth. Temperatures have fallen to a record −88°C.

The wind blows loose snow (above) into strange and sometimes beautiful shapes in Antarctica. Antarctica is an icy desert. The fierce winds create blizzards which are hard to endure.

Roald Amundsen was a Norwegian explorer. He travelled to the South Pole with sledges pulled by dogs. He beat Scott's party by over a month.

Richard Evelyn Byrd (left), an American airman and navigator, was the first man to fly over the South Pole. He achieved this feat in November, 1929. Byrd had already flown over the North Pole in 1926. Byrd made other contributions to polar exploration.

The Urge to Explore

Today, we live in the Space Age. Men have travelled in orbit around the Earth and, on 21 July 1969, the American astronaut Neil Armstrong became the first man to walk on the Moon's surface. However, despite all the great achievements in space, much still remains to be done in exploring our Earth.

Exploring the Land

In the last 40 years, a new method of mapping has become important. Maps are now made from photographs taken from aircraft. Because of this, there are maps of most of the world's land areas. However, this does not mean that all of the world has been properly explored. For example, in regions like the Himalayas, there are still peaks which no one has climbed.

Many densely forested lands await proper exploration and the world's deserts, whether hot like the Sahara or cold like Antarctica, still have secrets to yield. Every year, groups of trained people, including map-makers, biologists and geologists, make expeditions to remote places to collect new information. They often return with details of new animals and plants, deposits of minerals and many other interesting facts that they uncover.

Exploring the Oceans

The oceans cover more than seven-tenths of the Earth's surface. But, until recently, people were interested only in the surface of the oceans. Exploration of the oceans now involves the study of the sea-bed and the ocean waters, including living things at all levels.

The study of minerals on the sea-bed and the sources of food in seawater may be extremely important in the future. This is because the world's population is increasing quickly. Explorers of the oceans, therefore, have an important task to perform.

These ice ridges are in Antarctica, a continent where no one can live all the time. Exploration still goes on and scientists from many nations work there for periods.

Explorers' Who's Who

Alexander the Great (356-323 BC), king of Macedon, gained an empire which extended to India.
Amundsen, Roald (1872-1928), a Norwegian, was the first man to discover the North-West Passage and the first man at the South Pole.
Baffin, William (1584-1622), an Englishman, explored the Arctic.
Baker, Samuel (1821-93), a Briton, explored the Nile. He discovered Lake Albert.
Barents, Willem (1550-97), a Dutchman, explored the Arctic.
Bass, George (1763-1803?), a Briton, explored the coasts of south-east Australia and Tasmania.
Baudin, Nicolas (1750-1803), a Frenchman, explored the coasts of southern Australia.
Bering, Vitus (1681-1741), a Dane, explored the Bering Strait.
Bougainville, Louis Antoine de (1729-1811), the first Frenchman to sail around the world.
Burke, Robert O'Hara (1820-61), an Irishman, crossed Australia from south to north with William Wills (1834-61).
Burton, Richard F. (1821-90), a Briton, explored East Africa.
Byrd, Richard Evelyn (1888-1957), an American, was the first man to fly over the North and South poles.
Cabot, John (1450-98), an Italian, explored the Canadian coast.
Cabot, Sebastian (1474-1557), son of John Cabot, explored Canada and later the coast of South America.
Caillié, René (1799-1839), a Frenchman, explored the Sahara.
Cartier, Jacques (1491-1557), a Frenchman, explored eastern Canada.
Champlain, Samuel de (1567?-1635), a Frenchman, explored Canada.
Clapperton, Hugh (1788-1827), a Briton, explored Lake Chad.
Colombus, Christopher (1451-1506), an Italian, discovered the Americas for Spain.
Cook, James (1728-79), a Briton, explored the Pacific Ocean.
Coronado, Francisco Vasquez de (1510-54), a Spaniard, explored what is now the south-west USA.
Cortes, Hernando (1485-1547), a Spaniard, crushed the Aztec empire.
Dampier, William (1651-1715), a Briton, explored the coasts of north-western Australia.
Dias, Bartolomeu (1450-1500), a Portuguese seaman, sailed around the southern tip of Africa.
Drake, Sir Francis (1540?-96) led the first English round-the-world expedition.
Ericson, Leif (900s-1000s), a Viking, may have been the first European to land in North America.
Flinders, Matthew (1774-1814), a Briton, sailed around Australia, mapping the coastlines.
Franklin, John (1786-1847), a Briton, famous for his expeditions to the Arctic.
Frémont, John Charles (1813-90), an American, explored the western United States.
Frobisher, Martin (1535?-94), an Englishman, searched for a North-West Passage around Canada.
Fuchs, Vivian (1908-), a Briton, led an expedition which crossed Antarctica via the South Pole.
Gama, Vasco da (1469?-1524), a Portuguese navigator, found a sea route from Europe to India.
Grant, James (1827-92), a Scot, explored the Nile with John Speke.
Hanno, a Carthaginian, lived in about 450 BC. He explored the north-west coast of Africa.
Hartog, Dirk, a Dutchman born in the 1500s, was possibly the first European to set foot in Australia.
Henry the Navigator (1394-1460), a Portuguese prince, sponsored many voyages of discovery.
Herbert, Walter (Wally) (1934-), a Briton, led the first Trans-Arctic expedition.
Hillary, Edmund (1919-), a New Zealander, was the first man to reach the peak of Mount Everest.
Hudson, Henry (?-1611), an Englishman, sought a North-West Passage.
Hume, Hamilton (1797-1873), an Australian, explored the south-east of Australia.
Ibn Batuta (1304-77), an Arab, travelled in Africa and Asia.
Jansz, Willem (1570?-?), a Dutch seaman, was the first known European to sight Australia.
Joliet, Louis (1645-1700), a French-Canadian, explored the Mississippi river valley.
Laing, Alexander (1793-1826), a Scot, was the first European to see the African city of Timbuktu.
Lander, Richard (1803-34), a Briton, explored the River Niger with his brother John (1807-39).
La Salle, Sieur de, Robert Cavelier (1643-87), a Frenchman, explored in North America.
Leichhardt, Ludwig (1813-48), a German, explored north Australia.
Lewis, Meriwether (1774-1809), an American, explored the western USA with William Clark (1770-1838).
Livingstone, David (1813-73), a Scot, was the greatest explorer of the interior of Africa.
Mackenzie, Alexander (1764-1820), a Scot, explored north-west Canada.
Magellan, Ferdinand (1480-1521), a Portuguese navigator, led the first round-the-world voyage.
Marquette, Jacques (1637-75), a Frenchman, is known for his exploration of the Mississippi valley.
Nansen, Fridtjof (1861-1930), a Norwegian, explored the Arctic.
Nordenskjöld, Nils (1832-1901), a Swede, was the first man to sail through the North-East Passage.
Orellana, Francisco de (1500?-46?), a Spaniard, explored the Amazon.
Park, Mungo (1771-1806), A Scot, explored the River Niger.
Peary, Robert (1856-1920), an American, was the first man to reach the North Pole.
Pizarro, Francisco (1478?-1541), a Spaniard, defeated the Incas.
Polo, Marco (1254-1324), a Venetian. First European to explore central Asia and China.
Ross, James Clark (1800-62), a Scot, explored in both the Arctic and the Antarctic.
Scott, Robert Falcon (1868-1912), a Briton, explored in Antarctica.
Speke, John H. (1827-64), a Briton, discovered Lake Victoria.
Stanley, H. M. (1841-1904), who was born in Wales, explored in tropical Africa, notably the River Zaire.
Stuart, John McDouall (1815-66), a Scot, crossed Australia from south to north.
Sturt, Charles (1795-1869), a Briton, explored south-eastern and central Australia.
Tasman, Abel Janszoon (1603-59), a Dutchman, discovered Tasmania and New Zealand.
Vespucci, Amerigo (1451-1512), an Italian, explored the coasts of South America and the West Indies.
Willoughby, Hugh (1500-54), an Englishman, tried to find the North-West Passage.